Paediatric Cardiac Anaesthesia

Jutta Scheffczik

Paediatric Cardiac Anaesthesia

A Primer for the General Anaesthetist

 Springer

Jutta Scheffczik
Paediatric Cardiac Anaesthesia
Leeds General Infirmary
Leeds, UK

ISBN 978-3-031-90329-8   ISBN 978-3-031-90330-4   (eBook)
https://doi.org/10.1007/978-3-031-90330-4

© The Editor(s) (if applicable) and The Author(s), under exclusive license to Springer Nature Switzerland AG 2025

This work is subject to copyright. All rights are solely and exclusively licensed by the Publisher, whether the whole or part of the material is concerned, specifically the rights of translation, reprinting, reuse of illustrations, recitation, broadcasting, reproduction on microfilms or in any other physical way, and transmission or information storage and retrieval, electronic adaptation, computer software, or by similar or dissimilar methodology now known or hereafter developed.

The use of general descriptive names, registered names, trademarks, service marks, etc. in this publication does not imply, even in the absence of a specific statement, that such names are exempt from the relevant protective laws and regulations and therefore free for general use.

The publisher, the authors and the editors are safe to assume that the advice and information in this book are believed to be true and accurate at the date of publication. Neither the publisher nor the authors or the editors give a warranty, expressed or implied, with respect to the material contained herein or for any errors or omissions that may have been made. The publisher remains neutral with regard to jurisdictional claims in published maps and institutional affiliations.

This Springer imprint is published by the registered company Springer Nature Switzerland AG
The registered company address is: Gewerbestrasse 11, 6330 Cham, Switzerland

If disposing of this product, please recycle the paper.

Contents

1	**Anaesthetic Assessment and Planning**	1
	Additional Reading	5
2	**How to Read an Echo**	7
	Additional Reading	10
3	**Cardiovascular Changes at Birth**	11
	Additional Reading	13
4	**Pathophysiology and Anaesthetic Considerations**	15
	4.1 Regurgitant Lesions	15
	4.2 Obstructive Lesions	16
	4.3 Shunt Lesions	17
	4.4 Mixing Lesions	19
	4.5 Cyanosis and Hyperviscosity Syndrome	20
	Additional Reading	22

Part I Lesions

5	**Anomalous Left Coronary Artery from Pulmonary Artery**	25
	Additional Reading	27
6	**Aortic Arch Abnormalities: Vascular Rings**	29
	Additional Reading	32
7	**Aortic Arch, Interrupted**	33
	Additional Reading	36
8	**Aortic Valve**	37
	8.1 Aortic Valve Stenosis	37
	8.2 Aortic Valve Regurgitation	39
	Additional Reading	40

9	**Aorto-Pulmonary Window**	41
	Additional Reading	42
10	**Atrial Septal Defect**	43
	Additional Reading	45
11	**Arterio-Venous Malformation**	47
	11.1 Pulmonary Arterio-Venous Malformation	49
	11.2 Systemic Arterio-Venous Malformation	49
	11.3 Coronary Arterio-Venous Malformation	50
	Additional Reading	51
12	**Atrio-Ventricular Septal Defect, Complete**	53
	Additional Reading	56
13	**Atrio-Ventricular Septal Defect, Partial**	57
	Additional Reading	58
14	**Congenitally Corrected Transposition of the Great Arteries**	59
	Additional Reading	62
15	**Coarctation of the Aorta/Hypoplastic Aortic Arch**	63
	15.1 Coarctation of the Aorta	63
	15.2 Hypoplastic Aortic Arch	65
	Additional Reading	66
16	**Cor triatriatum**	67
	Additional Reading	69
17	**Double Inlet Left Ventricle**	71
	Additional Reading	75
18	**Double Outlet Right Ventricle**	77
	Additional Reading	80
19	**Ebstein's Anomaly/Malformation**	81
	Additional Reading	84
20	**Heterotaxy Syndrome: Left Atrial Isomerism, Polysplenia Syndrome**	85
	Additional Reading	88
21	**Heterotaxy Syndrome: Right Atrial Isomerism, Asplenia Syndrome**	89
	Additional Reading	92
22	**Hypoplastic Left Heart Syndrome**	93
	Additional Reading	95

23	**Mitral Valve**		97
	23.1	Mitral Valve Stenosis	97
	23.2	Systolic Anterior Motion of the Mitral Valve	98
	23.3	Mitral Valve Regurgitation	99
	23.4	Mitral Valve Prolapse	100
	Additional Reading		101
24	**Pulmonary Atresia with Intact Ventricular Septum**		103
	Additional Reading		105
25	**Partial Anomalous Pulmonary Venous Drainage**		107
	Additional Reading		109
26	**Patent Ductus Arteriosus**		111
	Additional Reading		112
27	**Pulmonary Valve**		113
	27.1	Pulmonary Valve Stenosis	113
	27.2	Pulmonary Regurgitation	115
	Additional Reading		116
28	**Tetralogy of Fallot's**		117
	Additional Reading		120
29	**Tetralogy of Fallot's with Pulmonary Atresia (Pulmonary Atresia with Ventricular Septal Defect)**		121
	Additional Reading		123
30	**Tricuspid Atresia (Hypoplastic Right Heart)**		125
	Additional Reading		127
31	**Tricuspid Valve**		129
	31.1	Tricuspid Stenosis	129
	31.2	Tricuspid Regurgitation	131
	Additional Reading		132
32	**Total Anomalous Pulmonary Venous Drainage**		133
	Additional Reading		137
33	**Transposition of the Great Arteries**		139
	Additional Reading		141
34	**Truncus Arteriosus**		143
	Additional Reading		146
35	**Ventricular Septal Defect**		147
	Additional Reading		149

Part II Operations

36 Central Shunts 153
  Additional Reading 156

37 Damus-Kaye-Stansel Operation 157
  Additional Reading 158

38 Glenn Shunt 159
  Additional Reading 161

39 Hybrid Procedure for Hypoplastic Left Heart Syndrome 163
  Additional Reading 165

40 Norwood Operation 167
  Additional Reading 169

41 Palliative Procedures 171
  41.1 Pulmonary Artery Band 171
  41.2 Balloon Atrial Septostomy (Rashkind Procedure) 172
  41.3 Ductal Stent 173
  Additional Reading 174

42 Aortic Valve Operations 175
  42.1 Ross Operation 175
  42.2 Konno Operation 176
  Additional Reading 177

43 Switch Operations 179
  43.1 Switch at Atrial Level: Mustard or Senning Procedure .. 181
  43.2 Switch at Ventricular Level: Rastelli, REV, Nikaidoh .. 181
  43.3 Switch at Arterial Level: Jatene Operation 182
  43.4 Double Switch for Congenitally Corrected TGA 183
  Additional Reading 184

44 Total Cavo-Pulmonary Connection (Fontan Operation) 185
  Additonal Reading 188

Appendix 189

Index 207

Abbreviations

ALCAPA	Anomalous left coronary artery (from) pulmonary artery
Ao	Aorta
AI	Aortic incompetence = regurgitation
AR	Aortic regurgitation
AS	Aortic stenosis
ASD	Atrial septal defect
AV	Atrioventricular or arterio-venous
AVSD	Atrioventricular septal defect
BAV	Bicuspid aortic valve
BTS	Blalock-Taussig Shunt, also known as Blalock-Taussig-Thomas shunt
BVF	Bi-ventricular function
CHD	Congenital heart disease
CHF	Congestive heart failure
CMP	Cardiomyopathy
CPB	Cardiopulmonary bypass
ccTGA	Congenitally corrected transposition of the great arteries
DILV	Double inlet left ventricle
DORV	Double outlet right ventricle
Echo	Echocardiography
FiO2	Inspired oxygen fraction
HCM	Hypertrophic cardiomyopathy
HLHS	Hypoplastic left heart
HOCM	Hypertrophic obstructive cardiomyopathy
IAA	Interrupted aortic arch
IVC	Inferior vena cava
IVS	Intact ventricular septum
LA	Left atrium
LCOS	Low cardiac output syndrome
LPA	Left pulmonary artery
L-R shunt	Left to right shunt
LV	Left ventricle or left ventricular

LVOT	Left ventricular outflow tract
LVOTO	Left ventricular outflow tract obstruction
MAPCA	Multiple aorto-pulmonary collateral arteries
MI	Myocardial infarction. Mitral incompetence = regurgitation
MPA	Main pulmonary artery
MR	Mitral regurgitation
MS	Mitral stenosis
NSAIDs	Nonsteroidal anti-inflammatory drugs
PA	Pulmonary artery or Pulmonary atresia
PAPVD	Partial anomalous pulmonary venous drainage
PAPVR	Partial anomalous pulmonary venous return
PBF	Pulmonary blood flow
pCO_2	Partial carbon dioxide measurement
PDA	Patent ductus arteriosus
PFO	Patent foramen ovale
PHT	Pulmonary hypertension
PI	Pulmonary incompetence = regurgitation
PR	Pulmonary regurgitation
PS	Pulmonary stenosis
PV	Pulmonary vein
PVD	Pulmonary vascular disease
PVR	Pulmonary vascular resistance
RA	Right atrium
R-L shunt	Right to left shunt
RPA	Right pulmonary artery
RV	Right ventricle or right ventricular
RVOT	Right ventricular outflow tract
RVOTO	Right ventricular outflow tract obstruction
SAM	Systolic anterior motion of the mitral valve
SCD	Sudden cardiac death
SVC	Superior vena cava
SVR	Systemic vascular resistance
SVT	Supraventricular tachycardia
TAPVD	Total anomalous pulmonary venous drainage
TAPVR	Total anomalous pulmonary venous return
TGA	Transposition of the great arteries
TOF	Tetralogy of Fallot
TR	Tricuspid regurgitation
TS	Tricuspid stenosis
VSD	Ventricular septal defect
VT	Ventricular tachycardia
WPW	Wolff-Parkinson-White syndrome or pre-excitation

Chapter 1
Anaesthetic Assessment and Planning

Abstract Preoperative assessment is essential for anaesthetic planning, deciding on the anaesthetic technique and risk management in cardiac children.

An anaesthetic assessment for a cardiac child includes a clinical examination (heart sounds/murmurs, auscultation of lungs, pulse status, as well as signs (or the lack thereof) of heart failure, etc), as well as an assessment of investigations. Chief amongst them is the Echocardiogram, which provides both an anatomical as well as a functional assessment. This is the main basis for the anaesthetic planning.

Keywords Anaesthetic assessment · Anaesthetic planning · Preoperative assessment · Preoperative investigations · Anaesthetic management · Paediatric cardiac anaesthesia

Anaesthetic preassessment is used for risk assessment, risk mitigation and gaining informed consent from the patient and parents.

Patients might have a singular cardiac problem (e.g. ventricular septal defect) and be otherwise healthy or have syndromes associated with cardiac lesions and therefore additional health issues which might influence the anaesthetic.

General Anaesthetic Assessment
- Previous anaesthetics: problems? Airway? Cardiovascular issues?
- Allergies
- Medication
- Fasting times
- Snoring/ Obstructive sleep apnoea: in children usually caused by large tonsils/ adenoids which might obstruct airway during induction and anaesthesia
- Wobbly/ loose teeth
- Cardiac lesion part of a syndrome with additional anaesthetic issues?

Specific Cardiac Assessment
- Echo: most important investigation for anaesthetic planning
    - Function
    - Details of lesion
    - Shunt blood flow
    - Severity of stenosis or regurgitation
    - Unexpected findings?
    - Coronary anatomy
    - Additional findings: e.g. tricuspid regurgitation: pulmonary hypertension?
- Observations:
    - heart rate, pulse status: pulses felt on all limbs?
    - blood pressure (ideally on more than one limb)
    - saturations (pre-and postductal in duct-dependent lesions)
    - respiratory rate, respiratory support, oxygen requirement
- ECG: sinus rhythm? QT interval? Blocks?
- Blood results: haemoglobin/ haematocrit, electrolytes (Potassium, Calcium, etc), coagulation tests
- Exercise test: provides a diagnostic and prognostic tool of cardiovascular fitness → needs a compliant patient, can only be assessed once walking/ running/ cycling
    - In babies: feeding well? Failure to thrive? Activity level?
    - In toddlers: activity level? Can keep up with siblings or playmates?
- Catheter laboratory results for pulmonary hypertension/ hyperoxia test:
    - Pulmonary hypertension: pulmonary artery pressures >20mmH
    - causes right ventricular hypertrophy, dilatation and failure
    - Causes functional tricuspid regurgitation
    - Risk of pulmonary hypertensive crisis
- Catheter lab: pulmonary vascular resistance PVR assessment: does oxygen decrease pulmonary artery pressure? → if not: moderate to high mortality (oxygen will not be able to decrease PA pressure in a pulmonary hypertensive crisis, leading to acute right heart failure)
- alternative vasodilators:
    - Nitric Oxide
    - Nebulised Milrinone, Prostaglandin or GTN
- Chest x-ray: might show cardiomegaly, plethoric lungs, etc.
- Clinical examination: Active, playing? Cyanosis? Clubbing? Pulse status? Heart murmur? Hepatomegaly?

1 Anaesthetic Assessment and Planning

Anaesthetic Planning Example
- Assess Echo for details of lesion (Fig. 1.1)
- Identify the pathological anatomy and draw the lesion
- Example: Tricuspid Atresia: draw normal heart anatomy and superimpose pathology (Fig. 1.2)
- Tricuspid atresia (complete obstruction)
- Atrial septal defect (R → L shunt)

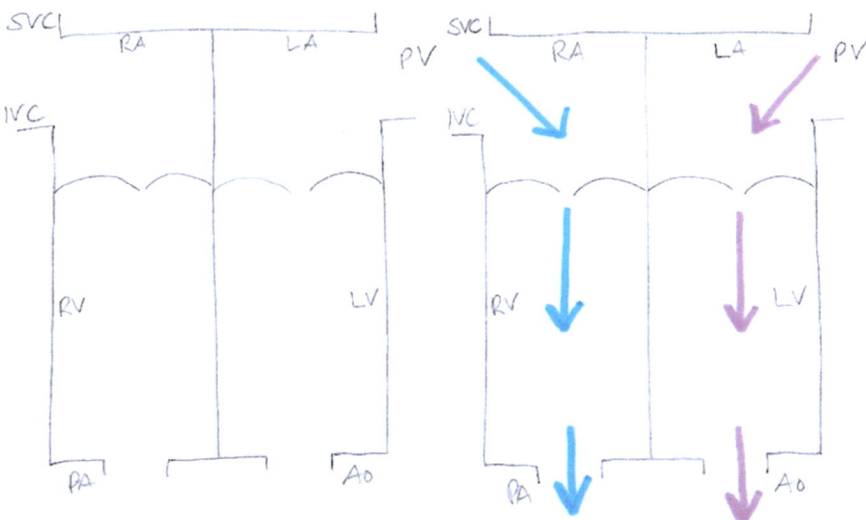

Fig. 1.1 Normal heart anatomy and blood flow

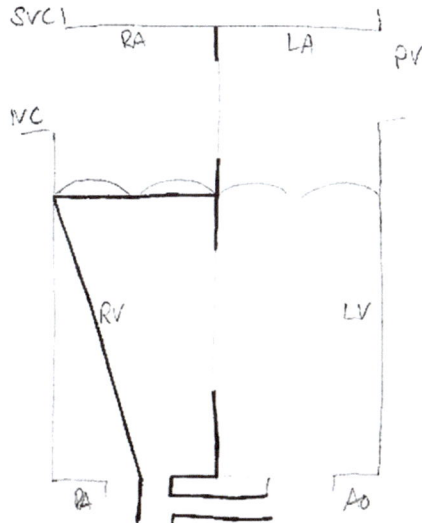

Fig. 1.2 Normal heart with tricuspid atresia, atrial septal defect, ventricular septal defect and duct superimposed on it

Fig. 1.3 Tricuspid atresia, anatomy and blood flow

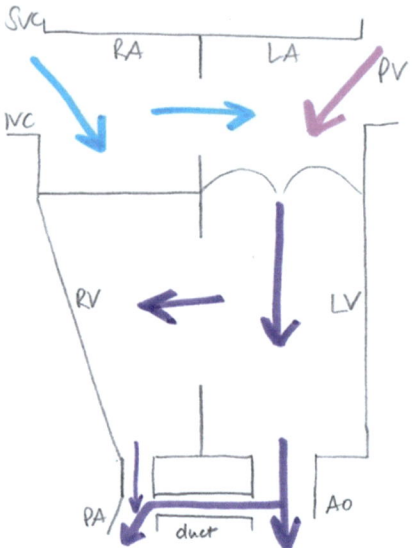

- Mixing of venous and arterial blood in left atrium
- Ventricular septal defect (L → R shunt)
- Small pulmonary valve and artery
- Duct dependent lesion (L → R shunt)

- Draw the blood flow (Fig. 1.3)
- Anaesthesia can't influence right and left atrial pressures
- Atrial septal defect needs to be unrestricted for systemic venous return
- Mixing is influenced by L → R shunt at ductal level
- Ventricular septal defect can't be influenced by anaesthetic
- L → R shunt at ductal level is influenced by pulmonary and systemic vascular resistance
- Keep saturations 80–85%
  - Keep pulmonary vascular resistance high with FiO2 21% and mild hypoventilation
  - Avoid increase in systemic blood pressure
- change as needed:
  - If saturation increases, decrease systemic vascular resistance
  - If saturation decreases, **either** decrease PVR with $FiO_2 > 21\%$ and mild hyperventilation **or** increase systemic vascular resistance SVR with a fluid bolus or vasopressors
- there is no evidence of one anaesthetic technique being superior

Additional Reading

1. Raviraj D, White M. Anaesthesia in children with congenital heart disease for non-cardiac surgery. Update in Anaesthesia. 2023;1:37.
2. Brown ML, Cradeur M, Staffa SJ, Nasr VG, Hernandez MR, DiNardo JA. Anaesthesia for non-cardiac surgery in children and young adults with Fontan physiology. Cardiol Young. 2023;33(10):1896–901.
3. Yamamoto T, Schindler E. Anaesthesia management for non-cardiac surgery in children with congenital heart disease. Anaesthesiol Intensive Ther. 2016;48(5):305–13.
4. Smith S, Walker A. Anaesthetic implications of congenital heart disease for children undergoing non-cardiac surgery. Anaesthesia Intensive Care Med. 2021;22(9):563–9.
5. Cannesson M, Earing MG, Collange V, Kersten JR, Riou B. Anesthesia for noncardiac surgery in adults with congenital heart disease. Anesthesiology. 2009;111(2):432–40.
6. Saettele AK, Christensen JL, Chilson KL, Murray DJ. Children with heart disease: risk stratification for non-cardiac surgery. J Clin Anesth. 2016;35:479–84.
7. Motiani P, Chhabra V, Ahmad Z, Sharma PK, Gupta A. Risk recognition and multidisciplinary approach for non-cardiac surgeries in paediatric cardiac patients: a retrospective observational study. Cureus [Internet]. 2020; [cited 2024 Apr 28]; Available from: https://www.cureus.com/articles/45693-risk-recognition-and-multidisciplinary-approach-for-non-cardiac-surgeries-in-paediatric-cardiac-patients-a-retrospective-observational-study
8. Wijesingha S, White M. Anaesthetic implications of congenital heart disease for children undergoing non-cardiac surgery. Anaesthesia Intensive Care Med. 2015;16(8):395–400.
9. Yuki K, Casta A, Uezono S. Anesthetic management of noncardiac surgery for patients with single ventricle physiology. J Anesth. 2011;25:247–56.
10. White MC, Peyton JM. Anaesthetic management of children with congenital heart disease for non-cardiac surgery. Continuing Educ Anaesthesia, Critical Care Pain. 2012;12(1):17–22.
11. Raviraj D, White M. Anaesthesia in children with congenital heart disease for non-cardiac surgery. Update in Anaesthesia. 2023;37.
12. Lei Lei E, Heggie J. Adult congenital heart disease and anesthesia: an educational review. Ramamoorthy C, editor. Pediatr Anesth. 2021;31(2):123–31.

Chapter 2
How to Read an Echo

Abstract Echocardiography is essential for planning an anaesthetic for a child with a cardiac condition. Knowing how to read and interpret an Echo helps with decisions about anaesthetic technique and monitoring.

An Echo provides both a detailed anatomical as well as a functional assessment. Stenotic lesion can be assessed in terms of gradient, regurgitant lesions in terms of the regurgitant fraction, shunt lesions can be assessed for amount of blood flow, direction of flow, as well as static and dynamic components.

Keywords Echocardiography · Heart function · Preoperative assessment · Preoperative investigations · Anaesthetic planning · Paediatric cardiac anaesthesia

An Echo is the most useful investigation for the planning of an anaesthetic as it shows the physiology and pathophysiology of the lesion, as well as assessment of function.

A Full Echo Report Will Contain the Following
- Patient identifier, date of exam, image quality, patient data such as indication, height and weight, body surface area
- Sequential segmental analysis of the entire heart
- Abdominal situs, cardiac position and apex orientation
- Venous return, atria and interatrial septum
- Atrioventricular connections and valves
- Ventricles and the interventricular septum
- Outflow tracts and ventriculoarterial connections
- Extra-pericardial great vessels

Sequential Segmental Analysis
- Situs solitus or inversus:
  - Situs inversus might impair ST segment interpretation in standard ECG electrode positioning
- Position of the heart: levo-, meso- or dextroposition: might influence ECG trace/reading, but only minimally
- Apex orientation: levo- or dextrocardia:
  - dextrocardia impairs ST segment interpretation in standard ECG electrode positioning
- Venous return to the heart:
  - superior vena cava important in anomalous return, such as partial or total pulmonary venous return or mixing lesions
- Atrial arrangement, defect(s) in the septum, membranes, etc.:
  - Size, position and number of defects, shunt volume, shunt direction
  - Position of membrane, gradient of stenosis, influence on atrial pressures or forward flow
  - Dilatation of right atrium increases risk of atrial arrhythmia
- Atrioventricular connections:
  - concordance: right atrium drains into right ventricle, left atrium drains into left ventricle
  - discordance (congenitally corrected transposition of the great arteries, double inlet left ventricle or heterotaxy syndromes)
- Atrioventricular valves: stenosis, regurgitation, dysplastic, single valve
  - obstruction is quantified by measuring the maximum velocity through the lesion, then calculating a pressure gradient with the modified Bernoulli equation: $\Delta P = 4\,V^2$
  - regurgitation: graded as mild, moderate, severe. Direction of regurgitant jet, regurgitant volume.
- Ventricular function: good/ impaired/ failure: for details see Functional assessment below.
- Ventricular septum (hypertrophy, defects, movement)
  - size, position and number of defects, shunt direction, shunt volume
  - muscle bundles/ stenosis of ventricular outflow tracts: gradient of stenosis (static, dynamic)
  - hypertrophy: symmetric or asymmetric: obstruction of blood flow?
- Ventriculo-arterial connections:
  - Concordance: pulmonary artery arises from right ventricle, aorta arises from left ventricle
  - Discordance (transposition of the great arteries, malposed great arteries)

- Ventriculo-arterial valves:
  - aortic and pulmonary valve: bicuspid aortic valve, stenosis, regurgitation, dysplasia, etc.
  - obstruction is quantified by measuring the maximum velocity through the lesion, then calculating a pressure gradient with the modified Bernoulli equation: $\Delta P = 4\ V^2$
  - Regurgitation: graded as mild, moderate, severe. Direction of regurgitant jet, regurgitant volume.
- Assessment of great vessels: size, form, course:
  - important for aortic arch abnormalities with compression of trachea and oesophagus, interrupted aortic arch, duct-dependent lesions, etc.
  - coronary anatomy if possible: origin, course and blood flow

Functional Assessment
Left ventricle:

- Systolic function: dependent on preload, diastolic function, RV function, intrathoracic and pericardial pressure
  - Fractional Shortening FS: percentage of change in LV short axis diameter, normal value: 30–45%
  - Ejection fraction: proportional change of LV volume during systole, expressed as a percentage: normal: 55–75%
  - Fractional area change FAC: proportional change in LV area during systole, expressed as a percentage: normal: 35–65%
  - Myocardial performance index MPI (Tei index): assesses overall function, both systolic and diastolic: sum of isovolumetric contraction time and isovolumetric relaxation time divided by the ejection time. Normal value: 0.37 +/− 0.05. MPI increases with impaired function.
- Diastolic function: impaired diastolic function causes elevated LV filling pressure. Left atrial size is marker of severity of diastolic dysfunction.
  - Usually not one marker, but several measurements of transmitral pulse wave doppler waveforms taken into account and commented as mild/ moderate/ severe diastolic dysfunction.

Right ventricle:

- Systolic function: similar to left ventricle, but challenging to assess due to ventricular shape and morphology.
  - Tricuspid Annular Plane Systolic Excursion: TAPSE. Age-dependent normal value, ranging from approximately 0.9 cm in neonates to 2.2–2.4 cm in adolescents.
  - RV fractional area change RVFAC, normal value: > 32%
- Diastolic function: measurement of tissue doppler, usually just commented on as mild/ moderate/ severe diastolic dysfunction

Additional Reading

1. Zawadka M, Marchel M, Andruszkiewicz P. Diastolic dysfunction of the left ventricle—a practical approach for an anaesthetist. Anaesthesiol Intensive Ther. 2020;52(3):237–44.
2. Budts W, Ravekes WJ, Danford DA, Kutty S. Diastolic heart failure in patients with the Fontan circulation: a review. JAMA Cardiol. 2020;5(5):590–7.
3. Grattan MJ, Mertens L. Echocardiographic assessment of ventricular function in pediatric patients: a comprehensive guide. Futur Cardiol. 2014;10(4):511–23.
4. Forshaw N, Broadhead M, Fenton M. How to interpret a paediatric echocardiography report. BJA Educ. 2020;20(8):278–86.
5. Bansal M, Sengupta PP. How to interpret an echocardiography report (for the non-imager)? Heart. 2017;103(21):1733–44.

Chapter 3
Cardiovascular Changes at Birth

Abstract Cardiovascular changes at birth occur separately, but interdependently. The main changes are lung aeration with decrease of pulmonary vascular resistance and increase in blood flow for gas exchange and provision of left ventricular preload, as well as the loss of placental gas exchange and increase of systemic vascular resistance by clamping the umbilical cord. Pressure changes in the atria cause the foramen ovale to close. Due to pulmonary and systemic vascular resistance changes there is flow reversal in the ductus arteriosus, which starts to close as well.

Keywords Antenatal circulation · Ductus arteriosus · Postnatal circulation · Cord clamping

The cardiovascular changes at birth aim to begin oxygenation of the blood in the lungs by unfolding and aerating the alveoli and close the shortcuts that diverted the blood from the pulmonary vasculature in utero (foramen ovale and ductus arteriosus).

- before birth (Fig. 3.1):
  - gas exchange via placenta
  - mixed blood (venous from systemic circulation and oxygenated blood from placenta) returns to right atrium
  - low pulmonary blood flow, majority of blood (about 90%) bypasses pulmonary circulation via 2 shunts:
    - intracardiac shunt: foramen ovale (right atrium to left atrium)
    - extracardiac shunt: ductus arteriosus Botalli (pulmonary artery to descending aorta)
  - pulmonary vascular resistance is high, so shunt flow is R → L
  - main source for preload of the left ventricle is umbilical venous return via the foramen ovale
  - lungs filled with liquid

Fig. 3.1 Antenatal circulation

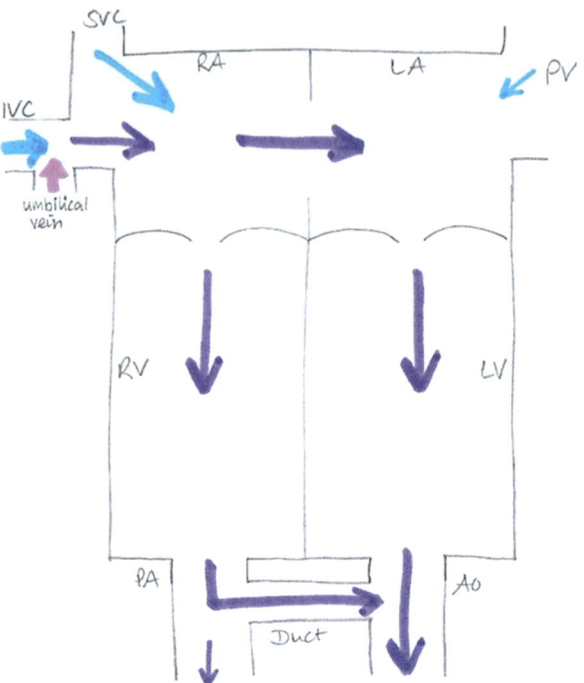

- during birth: beginning of airway fluid removal from lungs
  - absorption into lung tissue
  - loss of airway fluids due to thoracic compression during vaginal birth
- at/ immediately after birth: several mechanisms to occur separately, but interdependently over several minutes:
  - Lung aeration by taking breaths: gas exchange changes from placenta to lungs
    - Decrease in pulmonary vascular resistance → increase in pulmonary blood flow
    - Provision of preload to left heart from pulmonary venous return
    - Flow reversal in ductus arteriosus, then ductus starts to close
    - Transpulmonary pressures drive remaining airway fluid into interstitial tissue
  - Cord clamping:
    - Loss of umbilical venous return
    - Loss of preload to left atrium via the foramen ovale
    - Increase in systemic vascular resistance by removal of low-resistance placental circulation

Fig. 3.2 Circulation if cord is clamped before lung aeration

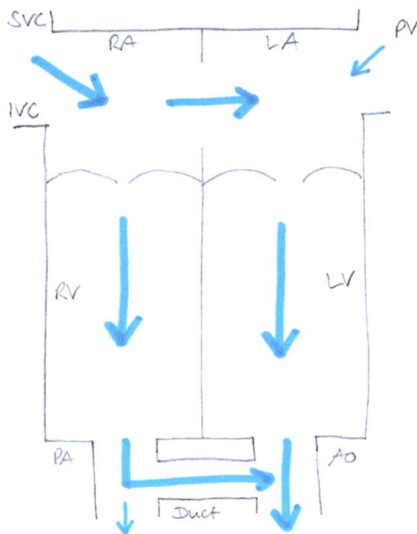

- if cord is clamped before lung aeration (Fig. 3.2):
  - Loss of placental gas exchange
  - No decrease in pulmonary vascular resistance
  - No increase in pulmonary blood flow
  - Decrease in cardiac output due to reduction in preload for the left ventricle and increased systemic vascular resistance
- After birth over several hours:
  - Airway fluid is slowly cleared from interstitial tissue, improving gas exchange
  - Ductus arteriosus closes

Additional Reading

1. Hooper SB, Te Pas AB, Lang J, Van Vonderen JJ, Roehr CC, Kluckow M, et al. Cardiovascular transition at birth: a physiological sequence. Pediatr Res. 2015;77(5):608–14.
2. Van Vonderen JJ, Roest AAW, Siew ML, Walther FJ, Hooper SB, Te Pas AB. Measuring physiological changes during the transition to life after birth. Neonatology. 2014;105(3):230–42.
3. Hooper SB, Roberts C, Dekker J, Te Pas AB. Issues in cardiopulmonary transition at birth. Semin Fetal Neonatal Med. 2019;24(6):101033.

Chapter 4
Pathophysiology and Anaesthetic Considerations

Abstract For anaesthetic planning the classification of lesions into the different pathophysiological principles is the most helpful. The four main pathologies are stenotic/obstructive lesions, regurgitant lesions, shunt lesions and mixing lesions. Complex cardiac disease can consist of more than one type of lesion and each needs to be assessed and planned for. An anaesthetist needs to be aware of the haemodynamic changes caused by anaesthetic drugs and techniques, how these changes influence the individual lesion and how to counteract them or change them when needed.

Keywords Cardiac classification, pathophysiology · Obstructive lesion · Shunt lesion · Regurgitant lesion · Mixing lesion, cyanosis, hyperviscosity syndrome · Anaesthetic planning, cyanotic heart lesion · Paediatric cardiac anaesthesia · Congenital cardiac lesion

There are several classification systems for congenital cardiac disease, such as cyanotic/noncyanotic, anatomy of structural defects or site of lesion, or physiological effects, such as on pulmonary blood flow or shunts.

From an anaesthetic point of view the physiological or pathophysiological classification makes most sense. There are four basic lesions, which can appear on their own or (in more complex cardiac lesions) as a combination of two or more.

4.1 Regurgitant Lesions

- Valvar

  - Dysplastic valves
  - Iatrogenic causes: balloon valvotomy, valve repair
  - Combination with other pathologies: Tricuspid regurgitation in pulmonary hypertension or valvar prolapse in ventricular septal defects

Right-Sided Lesions
- Pathophysiology: volume load of chamber proximal to regurgitant valve → initial muscular hypertrophy → late stage fibrosis and dilatation → dysfunction, failure
- Right atrial dilatation can cause arrhythmias
- Increased right atrial pressure can re-open a persistent foramen ovale PFO and create a R → L shunt
- Late stage right ventricular dysfunction causes backward failure with elevated central venous pressure, hepatomegaly, ascites, peripheral oedema, etc.

Left-Sided Lesions
- Pathophysiology: volume load of chamber proximal to regurgitant valve → initial muscular hypertrophy → late stage fibrosis and dilatation → dysfunction, failure
- Left ventricular forward failure: Low Cardiac Output Syndrome LCOS:
  - Reduced coronary perfusion
  - Systemic hypotension with reduced perfusion pressure
- Left ventricular backwards failure: pulmonary congestion with oedema and pulmonary hypertension

Anaesthetic Management
- avoid bradycardia
- decrease afterload to promote forward flow:
  - right-sided lesions: decrease pulmonary vascular resistance PVR: $FiO_2 > 21\%$, mild hyperventilation (pCO_2 high normal to slightly high)
  - left-sided lesions: decrease systemic vascular resistance SVR: anaesthetic drugs, vasodilators
- might need inotropes for impaired function
- be aware of arrhythmia with right atrial dilatation

4.2 Obstructive Lesions

- Valvar: Any valve can be stenotic, dysplastic or atretic
- Extra-valvar obstructions:
  - Muscular: infundibular, ventricular muscle bundles
  - Vascular: small blood vessels, Coarctation, hypoplastic or interrupted arch
  - Membranes: infra- or supra-valvar
- Complete obstructions (atretic valves, interrupted aortic arch) are duct-dependent for pulmonary blood flow (in right sided obstructions) or systemic blood flow (in left sided obstructions)

Pathophysiology of Right-Sided Lesions
- ↓ pulmonary blood flow
- Pressure increase proximal to the obstruction causes initial muscular hypertrophy, then dilatation, dysfunction and failure
- Late stage dilatation of valvar annulus proximal to the obstruction can cause regurgitation

Pathophysiology of Left-Sided Lesions
- ↓ systemic blood flow
- Low Cardiac Output Syndrome LCOS:
    - Reduced coronary perfusion
    - Systemic hypotension with reduced perfusion pressure
- Pressure increase proximal to the obstruction causes initial muscular hypertrophy, then dilatation, dysfunction and failure
- Late stage dilatation of valvar annulus proximal to the obstruction can cause regurgitation

Anaesthetic Management
- avoid tachycardia
- promote forward flow:
    - right-sided lesions: avoid increase in pulmonary vascular resistance PVR: FiO2 > 21%, mild hyperventilation
    - left-sided lesion: maintain systemic vascular resistance for adequate perfusion pressure for coronaries and tissue oxygenation
    - avoid a decrease in pulmonary vascular resistance: this can result in blood pooling in the lungs with plethoric lungs, oedema and hypertension.
- maintain or improve function with inotropes
- in duct-dependent lesions: maintain ductal patency:
    - adequate preload
    - Prostaglandin infusion
    - avoid non-steroidal anti-inflammatory drugs

4.3 Shunt Lesions

- A shunt is a connection between a low-pressure system and a high-pressure system
- Intracardiac shunts
    - Atrial septal defects ASD
    - Ventricular septal defects VSD
    - Atrio-ventricular defects AVSD

- Extracardiac shunts
    - Patent Ductus Arteriosus PDA
    - partial or total anomalous pulmonary venous drainage PAPVD/ TAPVD
    - Arterio-venous malformation AVM
- Shunt volume = blood going across or through the shunt = dependent on:
    - shunt size
    - *pressure difference* between the pressures on either side
- The greater the pressure difference, the more blood will shunt across
- Total shunt volume is limited by shunt size
- Pulmonary and systemic pressures can be manipulated with anaesthetic drugs and/or ventilation, therefore influencing shunt volume

    → challenging as pulmonary pressures can't be measured as continuously and reliably as systemic blood pressure
- Shunts are doubly dynamic:
    - Pressures on either side of shunt influence shunt volume
    - Shunt volume influences pressures on both sides of shunt
- Blood will shunt until pressures are equal, causing a bidirectional shunt or shunt reversal (Eisenmenger syndrome)

Pathophysiology of R → L shunt
- ↓ pulmonary blood flow
- Low systemic saturations = cyanosis
- Left ventricular volume load, leading to hypertrophy, then dilatation with fibrosis, dysfunction and failure

Pathophysiology of L → R Shunt
- ↓ systemic blood flow
- Large shunts can cause low cardiac output syndrome LCOS
- Large shunts can cause pulmonary hypertension
- Ventricular volume load, depending on shunt location either left ventricle or both
- an unrepaired or long-standing L → R shunt will lead to increased right heart pressures and shunt reversal: Eisenmenger syndrome, causing a R → L shunt with cyanosis

Anaesthetic Management
- maintain preload
- R → L shunts:
    - promote forward flow by decreasing pulmonary pressures PVR: FiO2 > 21%, mild hyperventilation
    - improve function with inotropes if needed

- L → R shunt:
  - Promote forward flow: avoid increasing systemic pressure SVR
  - Mild decrease in SVR might be helpful, maintain adequate perfusion pressures for coronaries and organ tissue perfusion
  - Improve function with inotropes if needed

4.4 Mixing Lesions

- Complex cyanotic lesions:
  - Hypoplastic left heart, tricuspid atresia, atrio-ventricular septal defects with unbalanced ventricles, large septal defects with equal pressures
  - Double inlet left ventricle DILV
  - Double outlet right ventricle DORV
  - Truncus arteriosus
  - Common atrium or common ventricle

Pathophysiology
- Mixing of venous and arterial blood
- Hypoxaemia, cyanosis
- Compensation for cyanosis leads to polycythaemia:
  - High haematocrit
  - High viscosity, can lead to hyperviscosity syndrome, if dehydrated
  - Coagulation problems

Anaesthetic Management
- Maintain preload
- Ideally keep pulmonary and systemic vascular resistance at pre-op level
- Adjust PVR and SVR depending on saturation and blood pressures:
- changes in pressures will direct blood flow to the pulmonary or systemic circulation, therefore influencing the mixture of venous and arterial blood
- saturations are determined by the mixing of venous and arterial blood and only indirectly influenced by oxygenation
- inspiratory oxygen can cause overcirculation of the lungs and relative hypovolaemia in the systemic circulation: use with caution!
- Example:
  - 50% venous blood/ 50% arterial blood create a saturation of low 80's
  - Decreasing pulmonary vascular resistance by increasing FiO2 directs more blood to the pulmonary circulation
    - 40% of blood is directed to the systemic circulation, returns deoxygenated, 60% is directed to the lungs, returns oxygenated
    - mixture of 40% venous blood/ 60% arterial blood → sats high 80's

- Decreasing PVR by further increasing FiO2 directs more blood to the pulmonary circulation
  - 0% of blood directed to the systemic circulation, 70% directed to the lungs
  - mixture of 30% venous blood/ 70% arterial blood → sats low 90's
  - if saturation is 100%: 0% venous blood/ 100% arterial blood: hypovolaemic systemic circulation

4.5 Cyanosis and Hyperviscosity Syndrome

Chronic hypoxia due to cyanotic heart lesions leads to changes in blood composition, causing coagulation abnormalities which can result in both an increased risk of bleeding and thromboembolic events.

- Secondary erythrocytosis is a physiological response to chronic tissue hypoxia in cyanotic heart disease via increased erythropoietin production
  - Improves oxygen transport capacity and tissue oxygenation
  - Leads to functional iron deficiency and microcytic red cells
  - Leads to elevated haemoglobin, haematocrit and red cell mass
    - increases blood viscosity and risk of thromboembolic events
    - optimal target haematocrit values for cyanotic heart disease are not known
- Secondary erythrocytosis/ increased haematocrit cause a variety of coagulation abnormalities, with an increase in bleeding risk:
  - Thrombocytopenia: platelet count and haematocrit are inversely related
  - Impaired fibrinogen function and accelerated fibrinolysis
  - Reduced levels of coagulation factors, mainly Vitamin K-dependent factors (II, V, VII, IX, X)
  - Some patients might have a low-grade disseminated intravascular coagulation DIC, a consumptive coagulopathy
    - point of care coagulation testing with Thrombelastography TEG or Rotation Thrombelastometry ROTEM can be more accurate than laboratory parameters
- Compensated erythrocytosis:
  - cyanosis causes release of erythropoietin
    - leads to increase in haematocrit and improved tissue oxygenation → erythropoietin levels decrease, haematocrit stays stable on a higher level, no symptoms of hyperviscosity

4.5 Cyanosis and Hyperviscosity Syndrome

- Decompensated erythrocytosis:
  - erythropoietin levels do not decrease after an increase in haematocrit
    - increases haematocrit and viscosity to harmful levels, causing decrease in tissue oxygenation and symptoms of hyperviscosity

Anaesthetic Management
- depending on surgery and expected blood loss
- Informed consent for clotting products and blood transfusion if necessary
- Transfusion of platelets, clotting factors, fibrinogen
- Antifibrinolytics: Tranexamic Acid or Epsilon Aminocaproic Acid EACA
- Platelet count and function increase within hours after phlebotomy (especially if haematocrit has been >65%)

Hyperviscosity Syndrome
- Viscosity is the measurement of internal resistance of a fluid to flow
- Hyperviscosity: blood is more resistant to flow
- Caused by
  - red cell deformity and rigidity
  - pathological elevation of both cellular and plasmatic blood components
  - dehydration
- also dependent on flow velocity, temperature, vessel size, endothelial function
- increased viscosity slows blood flow and paradoxically impairs microvascular circulation and tissue oxygenation, negating the beneficial effect of increased oxygen transport capacity
- Thrombosis risk increases with:
  - Iron deficiency
  - Maladaptive changes in endothelial function and structure
    - in cyanotic cardiac patients no correlation was found between raised haemoglobin or haematocrit and strokes/ thromboembolic events
- Symptoms dependent on iron deficiency and hydration status
- Clinical:
  - Cerebral problems: Headaches, visual disturbance, loss of concentration and mental acuity
  - Fatigue, muscle weakness, paraesthesia
  - Laboratory: high haemoglobin, high haematocrit, iron deficiency
  - Might have thromboembolic event: stroke, myocardial infarction, peripheral microthrombi
- Treatment/Anaesthetic management:
  - Multidisciplinary discussion with haematologist, cardiologist, involving surgical specialty, might need high-dependency or intensive care involvement
  - First line treatment is hydration: intravenous fluid bolus (discuss with haematologist: crystalloid or colloid, dependent on patient)

- Iron supplements for deficiency
- Phlebotomy is controversially discussed and seems rarely indicated in cyanotic heart disease:
    - should only be performed after adequate hydration and iron supplementation
    - needs replacement with equivalent volume of crystalloid or colloid

Additional Reading

1. Chowdhury D. Pathophysiology of congenital heart diseases. Ann Card Anaesth. 2007;10(1):19.
2. Gerrah R, Haller SJ, George I. Mechanical concepts applied in congenital heart disease and cardiac surgery. Ann Thorac Surg. 2017;103(6):2005–14.
3. Thiene G, Frescura C. Anatomical and pathophysiological classification of congenital heart disease. Cardiovasc Pathol. 2010;19(5):259–74.
4. Zabala LM, Guzzetta NA. Cyanotic congenital heart disease (CCHD): focus on hypoxemia, secondary erythrocytosis, and coagulation alterations. Hammer G, editor. Pediatr Anesth. 2015;25(10):981–9.
5. Tempe DK, Virmani S. Coagulation abnormalities in patients with cyanotic congenital heart disease. J Cardiothorac Vasc Anesth. 2002;16(6):752–65.

Part I
Lesions

Chapter 5
Anomalous Left Coronary Artery from Pulmonary Artery

Abstract An Anomalous Left Coronary Artery from the Pulmonary Artery (ALCAPA) is a rare coronary abnormality. Instead of oxygenated blood from a high-pressure system (Aorta) the coronary is perfused by deoxygenated blood from a low-pressure system (Pulmonary Artery). The pulmonary vascular resistance decreases in the weeks after birth, so while a neonate might be asymptomatic, myocardial ischaemia will occur at some point. Clinically the baby will show signs of distress, usually with feeding and the baby becoming pale/ grey and clammy due to anginal pain.

Anaesthetising a baby with ALCAPA is high risk and should have the availability of cardio-pulmonary bypass or ECMO as backup.

Keywords ALCAPA · Coronary abnormality · Left coronary artery, coronary steal · Angina, myocardial ischaemia · Vascular lesion · Paediatric cardiac anaesthesia · Congenital cardiac lesion

An anomalous left coronary artery arising from the pulmonary artery is a rare lesion with an approximate prevalence of 0.25–0.5%. This lesion is usually asymptomatic in the neonate and becomes apparent over time as the pulmonary vascular resistance decreases.

- Left coronary artery LCA comes from pulmonary artery instead of aorta
- LCA: deoxygenated blood from low-pressure system (Pulmonary Artery)
- RCA: oxygenated blood from high-pressure system (Aorta)

Pathophysiology
- neonate: high pulmonary vascular resistance PVR provides good left coronary blood flow, usually clinically asymptomatic (see Fig. 5.1)
- over time (>6 weeks): pulmonary vascular resistance decreases → reduced flow in left coronary artery, might develop into a retrograde steal into the pulmonary artery (see Fig. 5.2)

Fig. 5.1 ALCAPA immediately after birth with high pulmonary vascular resistance PVR: good forward flow into the left coronary artery LCA

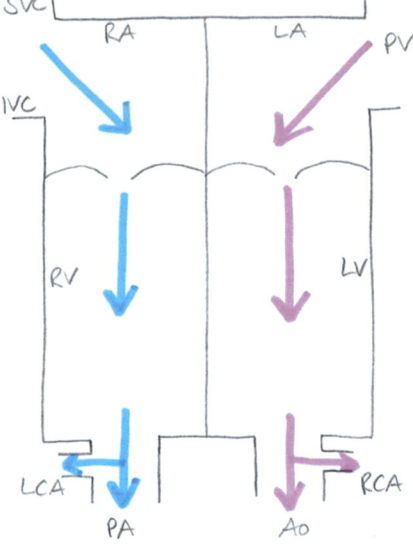

Fig. 5.2 ALCAPA blood flow with low PVR: reduced flow into left coronary artery, can lead to coronary steal into the pulmonary artery. This causes left ventricular ischaemia with rapidly progressive failure

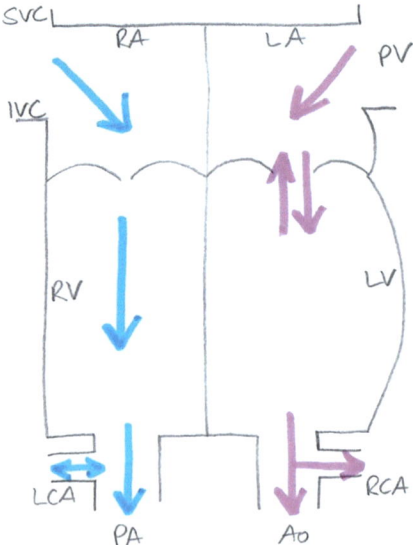

- main symptom is anginal pain due to myocardial ischaemia (signs of distress): pale, clammy, sweaty, especially with feeds
- collateral vessels might develop and provide adequate blood supply to the heart to survive into adulthood
- development of congestive heart failure due to ischaemia, left ventricular dilatation and failure, mitral regurgitation due to dilatation of the valve annulus, sudden cardiac death SCD

- Clinically:
    - Feeding issues due to angina/ left ventricular ischaemia
    - ECG: anterolateral ischaemia or infarction
    - Chest x-ray: normal or cardiomegaly in late stages
    - Echo: absent left coronary artery from aorta, impaired left ventricular function, left ventricular dilatation (late), mitral regurgitation (late), abnormal wall movement, signs of myocardial ischaemia
- **High risk anaesthesia**

Anaesthetic Management
- increase in myocardial O2 demand due to anaesthesia might not be met by left coronary artery → ST changes, bradycardia, ischaemic arrest
- keep myocardial oxygen demand low
    - Calm induction (premedication)
    - Avoid tachycardia
    - Keep systemic blood pressure at pre-op level
    - Keep pulmonary vascular resistance as high as possible: FiO2 21%, mild hypoventilation

Anaesthesia for Repaired Lesion
- Reimplantation of left coronary artery into aorta has excellent long-term outcomes
- Left ventricular function and mitral regurgitation usually improve to near-normal levels
    - check Echo for repair, left coronary flow, left ventricular function, mitral valve pathology

Additional Reading

1. Gupta K, Gupta M, Mehrotra M, Prasad J. Anaesthesia for repair of anomalous origin of left coronary artery from pulmonary artery. Indian J Anaesth. 2015;59(2):136–7.
2. Irvine DS, Rozava K, Theodotou A, Evans R, Huang J. Anesthesia consideration for a patient with incidentally diagnosed anomalous origin of right coronary artery originating from pulmonary trunk (ARCAPA): a case study. Cureus. 2022;14(9):e28955.
3. Beasley GS, Stephens EH, Backer CL, Joong A. Anomalous left coronary artery from the pulmonary artery (ALCAPA): a systematic review and historical perspective. Curr Pediatr Rep. 2019;7(2):45–52.
4. Lange R, Cleuziou J, Krane M, Ewert P, Pabst Von Ohain J, Beran E, et al. Long-term outcome after anomalous left coronary artery from the pulmonary artery repair: a 40-year single-Centre experience. Eur J Cardiothorac Surg. 2018;53(4):732–9.
5. Li D, Zhu Z, Zheng X, Wang Y, Wang Y, Xu R, et al. Surgical treatment of anomalous left coronary artery from pulmonary artery in an adult. Coron Artery Dis. 2015;26(8):723–5.

Chapter 6
Aortic Arch Abnormalities: Vascular Rings

Abstract There is a large variety of aortic arch abnormalities. Important for anaesthesia are complete vascular rings, encircling the trachea, bronchi and oesophagus, causing varying degrees of airway compression, vascular compression and feeding problems. There might be additional cardiac lesions or genetic abnormalities, but the main anaesthetic problem is airway compression and tracheomalacia, which require careful airway management. These children might need higher ventilation pressures with PEEP to overcome the airway compression. Vascular compression might influence pulmonary or upper limb perfusion, which has implications for the positioning of the arms.

Keywords Vascular lesion · Tracheal compression, tracheomalacia · Oesophageal compression, airway compression · Aortic arch abnormalities, aortic arch · Paediatric cardiac anaesthesia · Congenital cardiac lesion

Vascular rings are rare lesions with a prevalence of 1–2% in general population. The anatomy is very variable with different symptoms and degrees of severity depending on the individual lesion.

- Arch abnormalities include interrupted aortic arch (Chap. 7), arch hypoplasia (Chap. 15), coarctation (Chap. 15) and vascular rings
- Vascular rings are an encirclement of mediastinal structures like trachea, bronchi and oesophagus by the aorta and associated blood vessels (normal anatomy see Fig. 6.1)
- High degree of variability in anatomy and clinical presentation
- Complete: complete circle around trachea and oesophagus
    - double aortic arch (Fig. 6.2)
    - right aortic arch with left ligamentum arteriosus (Fig. 6.3)

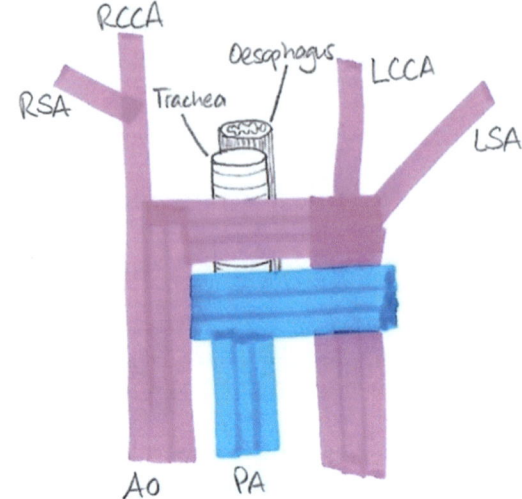

Fig. 6.1 Normal anatomy of the topography of aorta, pulmonary artery, trachea and oesophagus. *Ao* Aorta, *PA* pulmonary artery, *RSA* right subclavian artery, *RCCA* right common carotid artery, *LCCA* left common carotid artery, *LSA* left subclavian artery

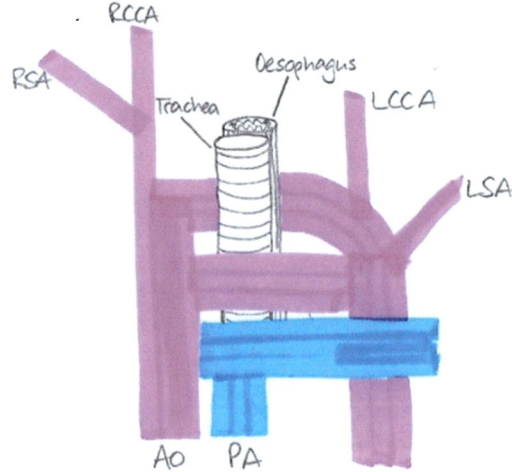

Fig. 6.2 Double aortic arch, complete vascular ring. *Ao* Aorta, *PA* pulmonary artery, *RSA* right subclavian artery, *RCCA* right common carotid artery, *LCCA* left common carotid artery, *LSA* left subclavian artery

- Incomplete: no complete circle, but compression to various degrees:
  - anomalous innominate artery
  - anomalous left pulmonary artery (vascular sling, Fig. 6.4)
  - aberrant right subclavian artery
- Clinical: depending on individual lesion
  - Asymptomatic
  - Airway obstruction: stridor, wheeze, cough, recurrent chest infections
  - Oesophageal obstruction: dysphagia, feeding problems, aspiration with aspiration pneumonia

Fig. 6.3 Right aortic arch with ligamentum arteriosus. *Ao* Aorta, *PA* pulmonary artery, *RSA* right subclavian artery, *RCCA* right common carotid artery, *LCCA* left common carotid artery, *LSA* left subclavian artery

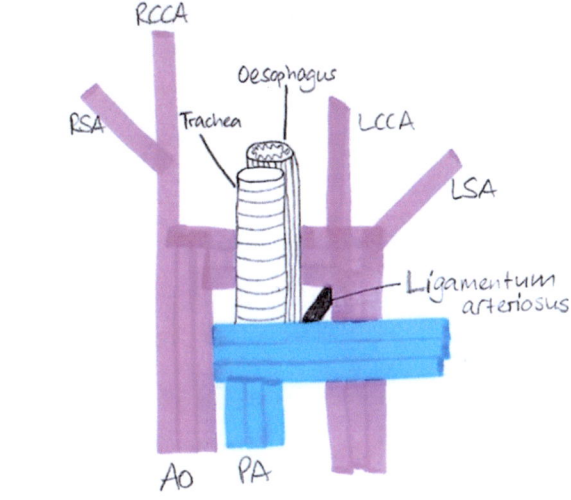

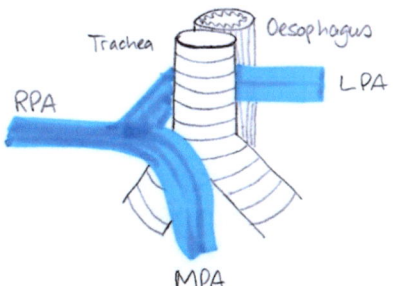

Fig. 6.4 Vascular sling: the left pulmonary artery arises from the right pulmonary artery and runs in between the oesophagus and the trachea, which can lead to compression of all structures, trachea, oesophagus and left pulmonary artery. *MPA* main pulmonary artery, *RPA* right pulmonary artery, *LPA* left pulmonary artery

  – Vascular compression or obstruction
  – Failure to thrive
- can be associated with other cardiac lesions or genetic abnormalities
- in isolated lesions: no cardiac precautions necessary due to normal heart function

Anaesthetic Management
- tracheal obstruction and tracheomalacia: careful airway management
- might need higher ventilation pressures and PEEP to overcome airway compression
- careful positioning of arm in subclavian steal, ensure good perfusion, saturation probe on left hand
- in vascular slings: possible compression/ stenosis of left pulmonary artery → reduced perfusion of left lung: can't be influenced by anaesthesia
- additional cardiac lesions?

Anaesthesia in Repaired Lesion
- usually good repair
- symptoms might persist
- might have persistent tracheomalacia
- Check Echo if additional cardiac lesions

Additional Reading

1. Das S, Aggarwal S. Airway and esophageal compression from double aortic arch in a case of pentalogy of Fallot: Anesthetic management. Indian Anaesth Forum. 2017;18(2):82.
2. Kussman BD, Geva T, McGowan FX. Cardiovascular causes of airway compression. Pediatr Anesth. 2004;14(1):60–74.
3. Priya S, Thomas R, Nagpal P, Sharma A, Steigner M. Congenital anomalies of the aortic arch. Cardiovasc Diagn Ther. 2018;8(Suppl 1):S26–44.
4. Wadle M, Joffe D, Backer C, Ross F. Perioperative and Anesthetic considerations in vascular rings and slings. Semin Cardiothorac Vasc Anesth. 2024;
5. Rato J, Zidere V, François K, Boon M, Depypere A, Simpson JM, et al. Post-operative outcomes for vascular rings: a systematic review and meta-analysis. J Pediatr Surg. 2023;58(9):1744–53.
6. Friedman K. Preoperative physiology, imaging, and management of interrupted aortic arch. Semin Cardiothorac Vasc Anesth. 2018;22(3):265–9.

Chapter 7
Aortic Arch, Interrupted

Abstract An interrupted aortic arch is a rare, duct-dependent lesion, where the perfusion for the lower limbs and sometimes the left arm is provided from a R → L shunt via the duct. Clinical signs depend on the location of the interruption, with cyanosis, low blood pressure and weak pulses in the lower limbs. Anaesthetic management includes a Prostin infusion to keep the duct open and to keep the pulmonary blood pressure high with a low FiO2 and mild hypoventilation, to keep an adequate perfusion pressure for the kidneys and gut.

Keywords Aortic arch, aortic arch abnormalities · Vascular lesion · Obstructive lesions · R->L shunt · Duct-dependent · Paediatric cardiac anaesthesia · Congenital cardiac lesion

An interrupted aortic arch is a rare lesion, about 1% of congenital heart disease (an extreme form of Coarctation or hypoplastic aortic arch). This is a duct-dependent lesion, where the blood flow to the abdomen and lower limbs comes from the pulmonary artery, resulting in hypotension and cyanosis of the abdomen and lower limbs.

The classification is an anatomical one, depending on the site of interruption.

- 3 types
  - A: 35% (Figs. 7.1 and 7.2)
  - B: 45% (Figs. 7.1 and 7.3)
  - C: 20% (Figs. 7.1 and 7.4)

- Associated with DiGeorge syndrome, ventricular septal defect VSD, left ventricular outflow tract abnormalities, bicuspid aortic valve BAV, etc.

7 Aortic Arch, Interrupted

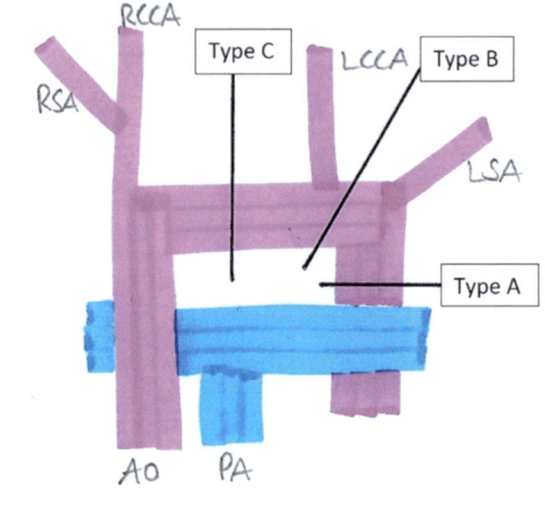

Fig. 7.1 Overview of all types of interrupted aortic arches. *Ao* Aorta, *PA* pulmonary artery, *RSA* right subclavian artery, *RCCA* right common carotid artery, *LCCA* left common carotid artery, *LSA* left subclavian artery

Fig. 7.2 Interrupted Aortic Arch type A: *Ao* Aorta, *MPA*: main pulmonary artery, *RPA* right pulmonary artery, *LPA* left pulmonary artery, *RSA* right subclavian artery, *RCCA* right common carotid artery, *LCCA* left common carotid artery, *LSA* left subclavian artery
Saturation and blood pressure:
Right arm: normal
Left arm: normal
Legs: low

- Clinically:
    - Blood pressure and saturation differences between upper and lower limbs
    - Lower limb cyanosis, weak pulses
    - Respiratory distress, developing heart failure
    - Chest x-ray/ ECG: non-specific
    - Echo: shows interrupted aortic arch and associated lesions

7 Aortic Arch, Interrupted

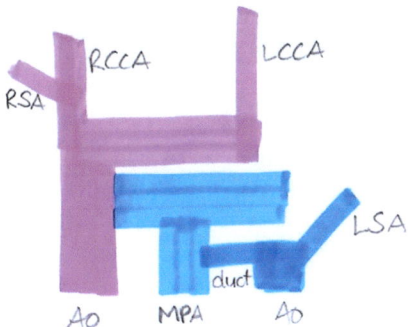

Fig. 7.3 Interrupted Aortic Arch type B: *Ao* Aorta, *MPA* main pulmonary artery, *RSA* right subclavian artery, *RCCA* right common carotid artery, *LCCA* left common carotid artery, *LSA* left subclavian artery
Saturation and blood pressure:
Right arm: normal
Left arm: low
Legs: low

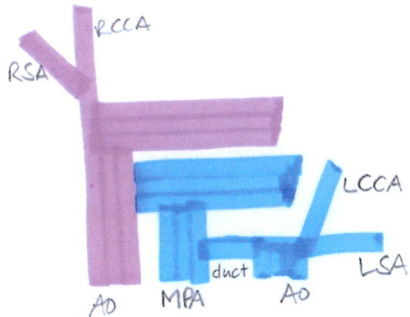

Fig. 7.4 Interrupted Aortic Arch type C: *Ao* Aorta, *MPA* main pulmonary artery, *RSA* right subclavian artery, *RCCA* right common carotid artery, *LCCA* left common carotid artery, *LSA* left subclavian artery
Saturation and blood pressure:
Right arm: normal
Left arm: low
Legs: low

Anaesthetic Management
- duct-dependent → Prostaglandin infusion
- Pulmonary blood pressure = systemic blood pressure for lower body: kidney and gut perfusion might be compromised
    - keep pulmonary blood pressure as high as possible: FiO2 21% and mild hypoventilation

- Careful fluid management, maintain preload
- might need vasopressors or inotropes
- cave associated lesions:
    - Ventricular septal defect: L → R shunt
    - Bicuspid aortic valve: aortic stenosis?
    - Truncus arteriosus: mixing lesion
    - DiGeorge

Anaesthetic for Repaired Lesion
- Usually good repair
- Stenosis at anastomosis site possible
- Residual aortic stenosis with bicuspid aortic valve or sub-valvular aortic stenosis
- Ventricular septal defect closed? Residual defect?
- Arrhythmia post-closure?

Additional Reading

1. Brown JW, Ruzmetov M, Okada Y, Vijay P, Rodefeld MD, Turrentine MW. Outcomes in patients with interrupted aortic arch and associated anomalies: a 20-year experience. Eur J Cardiothorac Surg. 2006;29(5):666–73.
2. Burbano-Vera N, Zaleski KL, Latham GJ, Nasr VG. Perioperative and Anesthetic considerations in interrupted aortic arch. Semin Cardiothorac Vasc Anesth. 2018;22(3):270–7.
3. Friedman K. Preoperative physiology, imaging, and management of interrupted aortic arch. Semin Cardiothorac Vasc Anesth. 2018;22(3):265–9.

Chapter 8
Aortic Valve

Abstract This chapter describes aortic valve lesions. Aortic stenosis can be congenital or acquired, with clinical symptoms dependent on the degree of stenosis. Mild stenosis is asymptomatic, moderate to severe stenosis causes high ventricular pressures with low cardiac output syndrome. Anaesthetic management aims to avoid tachycardia while maintaining adequate blood pressure for coronary perfusion. Inotropes might be needed, depending on ventricular function.

Aortic regurgitation is rare as an isolated lesion. Symptoms depend on severity of regurgitation and associated lesions. Anaesthetic management aims to promote forward flow by carefully reducing systemic vascular resistance while maintaining coronary perfusion pressures and to avoid bradycardia.

Keywords Aortic valve, valvular lesions, valve lesion · Obstructive lesion, regurgitant lesion · Duct-dependent · Paediatric cardiac anaesthesia · Congenital cardiac lesion

Congenital valvar abnormalities in isolation are rare, they are usually associated with other cardiac lesions. Even when clinically asymptomatic, these lesions cause altered haemodynamics, which makes the valve prone to calcification and has been associated with aortic dilatation.

8.1 Aortic Valve Stenosis

- 3–6% of congenital heart disease, M:F = 4:1
- Types:
    - valvular: most common: bicuspid aortic valve BAV
    - supra-valvular: most common: Williams syndrome
    - sub-valvular (membrane): Shone's complex

- Pathophysiology: left ventricular hypertrophy and ventricular pressures → impaired ventricular function, dilatation → ischaemia → arrhythmia → sudden cardiac death SCD
- Severe or critical aortic stenosis in neonates: duct-dependent, needing urgent balloon valvotomy (Rashkind procedure, Chap. 42)
- Clinical:
  - Hypotension, low cardiac output syndrome LCOS
  - Low exercise tolerance, feeding issues, failure to thrive
  - Murmur
  - ECG: might be normal, might show LV hypertrophy or ischaemia
  - Chest x-ray: might have cardiomegaly with congestive heart failure
  - Echo: details of lesion, ventricular function, ventricular hypertrophy?, coronary artery involvement?, gradient (calculated with the modified Bernoulli equation with V as maximum velocity: $\Delta P = 4\ V^2$)
    - mild: <40 mmHg
    - moderate: 40–70 mmHg
    - severe: >70 mmHg
- **High risk anaesthesia for severe stenosis or associated syndromes**

Anaesthetic Management
- avoid decrease in systemic blood pressure: fluids, vasopressors
- avoid decrease in pulmonary blood pressure: FiO2 21%, mild hypoventilation
- avoid tachycardia
- maintain preload
- maintain coronary perfusion pressures
- maintain ductal patency when applicable:
  - Prostaglandin infusion
  - L → R shunt:
    - avoid decrease in pulmonary vascular resistance: FiO2 21%, mild hypoventilation
    - avoid increase in systemic vascular resistance: good anaesthesia and analgesia

Anaesthesia for Repaired Lesion
- Repair:
  - Ballooning of valve (balloon valvotomy) → might cause aortic regurgitation, re-stenosis might occur
  - Konno operation: Aortic root replacement with ventricular septum extension (Chap. 42)
  - Ross operation: Pulmonary valve replaces aortic valve; homograft, bioprosthetic or mechanical pulmonary valve for pulmonary valve replacement (Chap. 42)

- Check Echo for:
  - Residual stenosis or re-stenosis? Gradient?
  - Coronary perfusion
  - Ventricular function (impairment might persist)
  - Additional lesions?
  - Aortic dilatation
  - Replacement valve function: calcification? Structural valve deterioration?
- Post-Konno: arrhythmia?
- Check for anticoagulation medication in valve replacement:
  - Homograft/ bio-prosthesis: might have Aspirin for coagulation management
  - Mechanical valve: warfarin for coagulation management
- Check local/national guidelines regarding Endocarditis prophylaxis

8.2 Aortic Valve Regurgitation

- Rare as an isolated lesion
- Associated with sub-valvular aortic stenosis and Rheumatic Fever
- Pathophysiology:
  - left ventricular volume load→ dilatation→ dysfunction, impairment
  - Pulmonary hypertension PHT due to backward failure
- Clinical:
  - Low cardiac output syndrome LCOS, angina, syncope, low exercise tolerance, congestive heart failure
  - Murmur
  - ECG: Left ventricular hypertrophy in severe cases
  - Chest x-ray: cardiomegaly
  - Echo: valve regurgitation, left ventricular hypertrophy +/− dilatation, left ventricular function, pulmonary hypertension PHT
- **High risk anaesthesia for severe regurgitation**

Anaesthetic Management
- Decrease left ventricular afterload by decreasing systemic vascular resistance (good anaesthetic)
- Avoid bradycardia
- Might need inotropes, depending on left ventricular function
- Manage pulmonary hypertension by decreasing pulmonary vascular resistance: oxygen, mild hyperventilation

Anaesthesia for Repaired Lesion
- Valve replacement: check Echo for
  - residual regurgitation
  - coronary blood flow
  - Replacement valve function: calcification? Structural valve deterioration?
  - Contractility/ function
  - Additional/ associated lesions
- Anticoagulation medication:
  - Homograft/ bio-prosthesis: might have Aspirin for coagulation management
  - Mechanical valve: usually warfarin for coagulation management
- Left ventricular dysfunction might be irreversible
- Pulmonary hypertension might be irreversible
- Check local/national guidelines regarding Endocarditis prophylaxis

Additional Reading

1. Paul A, Das S. Valvular heart disease and anaesthesia. Indian J Anaesth. 2017;61(9):721.
2. Balmer C. Balloon aortic valvoplasty in paediatric patients: progressive aortic regurgitation is common. Heart. 2004;90(1):77–81.
3. Burch TM, McGowan FX, Kussman BD, Powell AJ, DiNardo JA. Congenital Supravalvular aortic stenosis and sudden death associated with Anesthesia: what's the mystery? Anesth Analg. 2008;107(6):1848–54.
4. Etnel JRG, Takkenberg JJM, Spaans LG, Bogers AJJC, Helbing WA. Paediatric subvalvular aortic stenosis: a systematic review and meta-analysis of natural history and surgical outcome. Eur J Cardiothorac Surg. 2015;48(2):212–20.
5. Gupta P, Tobias J, Goyal S, Miller M, Melendez E, Noviski N, et al. Sudden cardiac death under anesthesia in pediatric patient with Williams syndrome: a case report and review of literature. Ann Card Anaesth. 2010;13(1):44.
6. Ross F, Everhart K, Latham G, Joffe D. Perioperative and Anesthetic considerations in Pediatric valvar and subvalvar aortic stenosis. Semin Cardiothorac Vasc Anesth. 2023;27(4):292–304.
7. Schlein J, Wollmann F, Kaider A, Wiedemann D, Gabriel H, Hornykewycz S, et al. Long-term outcomes after surgical repair of subvalvular aortic stenosis in pediatric patients. Front Cardiovasc Med. 2022;9:1033312.
8. Singh GK. Congenital aortic valve stenosis. Children. 2019;6(5):69.
9. Zaidi M, Premkumar G, Naqvi R, Khashkhusha A, Aslam Z, Ali A, et al. Aortic valve surgery: management and outcomes in the paediatric population. Eur J Pediatr. 2021;180(10):3129–39.
10. Watabe A, Saito H, Harasawa K, Morimoto Y. Anesthetic management for severe aortic regurgitation in an infant repaired by Ross procedure. J Anesth. 2009;23(2):270–4.

Chapter 9
Aorto-Pulmonary Window

Abstract An aortopulmonary window is a L → R shunt lesion between the ascending aorta and the pulmonary artery, with symptoms depending on the size of the shunt. The size of the window determines how quickly high output heart failure develops. The decrease in pulmonary vascular resistance in the first weeks of life increases the shunt volume and the pulmonary blood flow increases. Anaesthetic management aims to keep the shunt volume as low as possible with high pulmonary vascular resistance and a reasonably low systemic blood pressure.

Keywords Aortopulmonary window · L → R shunt · Vascular lesion · Pulmonary overcirculation · Pulmonary hypertension · Paediatric cardiac anaesthesia · Congenital cardiac lesion

An aorto-pulmonary window, also known as aorto-pulmonary septal defect, is a rare defect with a L → R shunt between the systemic and pulmonary circulation. The symptoms are dependent on the size of the shunt and the pressure difference between the two circulations, complications arise from the overcirculation of the pulmonary vasculature, leading to pulmonary hypertension and progressive congestive heart failure.

- defect between the ascending aorta and the main pulmonary artery (failure to septate and completely divide the embryonic truncus, Fig. 9.1):
  – Antenatal: R → L shunt due to high pulmonary vascular resistance
  – Postnatal: L → R shunt due to fall in pulmonary vascular resistance
    - decreasing pulmonary vascular resistance over the first weeks of life increases shunt
- Difference between A-P window and Truncus arteriosus: Truncus has one valve, A-P window has two valves

Fig. 9.1 Aorto-pulmonary window with L → R shunt

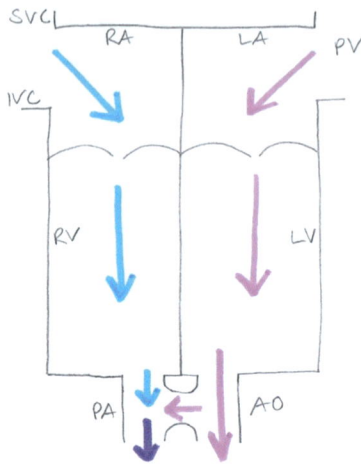

- Clinically:
  - Dyspnoea, Tachypnoea due to high pulmonary blood flow
  - Developing congestive heart failure and pulmonary hypertension
  - Murmur: systolic ejection
  - ECG: normal or left ventricular hypertrophy, biventricular hypertrophy
  - Chest x-ray: normal or cardiomegaly, plethoric lungs
  - Echo: size of lesion, ventricular function, associated defects such as coronary abnormalities

Anaesthetic Management
- L → R shunt:
  - Avoid decrease in pulmonary vascular resistance: FiO2 21%, mild hypoventilation
  - Avoid increase in systemic vascular resistance: good anaesthesia and analgesia
- Patients in congestive heart failure might need inotropes

Anaesthesia for Repaired Lesion
- Usually good repair, no anaesthetic issues
- Check Echo for

  - Stenosis at patch site
  - Residual pulmonary hypertension? (tricuspid regurgitation?)
  - Contractility/function
  - Coronary anatomy and blood flow
  - Associated lesions?

Additional Reading

1. Kulkarni M, Sushma TK, Shastry A. Anaesthesia in a child with uncorrected aortopulmonary window for non-cardiac surgery. Sri Lankan J Anaesthesiol. 2020;28(2):153–5.
2. Dharmalingam S, Pillai R, Karuppiah S, Sahajanandan R, George G. Case report of aortopulmonary window with undiagnosed interrupted aortic arch: role of transesophageal echocardiography. Ann Card Anaesth. 2016;19(1):152–3.

Chapter 10
Atrial Septal Defect

Abstract Atrial septal defects are L → R shunt lesions, with symptoms depending on the size of the shunt. Small shunts might be asymptomatic and remain undetected until adult age. Long-standing or large shunts can cause congestive heart failure, pulmonary hypertension and shunt reversal (Eisenmenger's syndrome). The influence of anaesthesia on atrial pressures to change shunt volume is limited. Anaesthetic management therefore is to maintain preload and manage complications such as congestive heart failure or pulmonary hypertension. Large shunt volumes can cause right atrial dilatation which can lead to arrhythmias.

Keywords Atrial septal defect · Shunt lesion, shunt reversal, Eisenmenger's syndrome · Arrhythmia, pulmonary hypertension · L → R shunt · Paediatric cardiac anaesthesia · Congenital cardiac lesion

Atrial septal defects make up about 5–10% of congenital heart defects and are more prevalent in females (M:F = 1:2). Small shunts might go undetected for decades until late stage complications arise from the continuous overcirculation of the pulmonary vasculature and subsequent right heart impairment and failure.

- Types:
  - Primum ≈30%
  - Secundum ≈50–70%
  - Sinus venosus (unroofed coronary sinus) ≈ 10%
- Clinical:
  - Symptoms depending on shunt volume, asymptomatic in small lesions (Fig. 10.1)
  - Heart murmur
  - ECG: might be normal, right ventricular hypertrophy, right bundle branch block RBBB
  - Chest x-ray: might be normal, cardiomegaly in late stages
  - Echo: lesion, volume loaded right atrium and ventricle

Fig. 10.1 Atrial Septal Defect ASD with L → R shunt

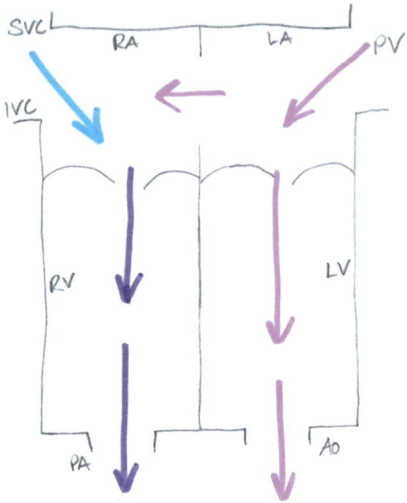

- Late stages: Right ventricular dilatation and progressive dysfunction

  - Right atrial dilatation, atrial arrhythmia
  - Pulmonary hypertension (with tricuspid regurgitation)
  - Eisenmenger syndrome (shunt reversal)

Anaesthetic Considerations
- L → R shunt
- Maintain preload
- Avoid decrease in pulmonary vascular resistance: FiO2 21%, mild hypoventilation
- Problems with late stages (Fig. 10.2):

  - Dilatation of right atrium: arrhythmias
  - Dilatation of right ventricle: impaired function
  - Stroke volume can no longer be improved by preload
  - Increase in right-sided pressures: reduction in shunt volume → can cause shunt reversal with cyanosis (Eisenmenger's syndrome)
  - Increase in pulmonary pressures due to high volume load: pulmonary hypertension and tricuspid regurgitation

Anaesthesia for Repaired Lesion
- Surgical repair or device closure
- Usually no precautions necessary
- Echo:

  - Residual shunt?
  - Persistent pulmonary hypertension?
  - Persistent dysfunction?

Fig. 10.2 Atrial Septal Defect, late stage with right heart dilatation, equal atrial pressures with bidirectional shunt and tricuspid regurgitation

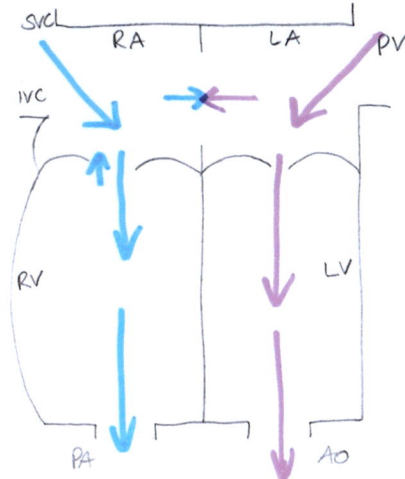

Additional Reading

1. Calvert PA, Klein AA. Anaesthesia for percutaneous closure of atrial septal defects. BJA Educ. 2008;8(1):16–20.
2. Madaan V, Gupta R. Anaesthetic management of a case of large ASD with severe pulmonary hypertension – case presentation. Ain-Shams J Anesthesiol. 2022;14(1):32.
3. Park Y, Kim J. Anesthetic management of a patient with large atrial septal defect undergoing laparoscopic cholecystectomy: a case report. Saudi J Anaesth. 2020;14(2):249–52.
4. Alkashkari W, Albugami S, Hijazi ZM. Current practice in atrial septal defect occlusion in children and adults. Expert Rev. Cardiovasc Ther. 2020;18(6):315–29.
5. Backer CL, Eltayeb O, Mongé MC, Mazwi ML, Costello JM. Shunt lesions part I: patent ductus arteriosus, atrial septal defect, ventricular septal defect, and atrioventricular septal defect. Pediatr Crit Care Med. 2016;17(8):S302–9.
6. Suryavanshi K, Deshpande J. Anaesthesia management of case of atrial septal defect (cardiac disease) for proximal humerus fracture (non-cardiac case). Arch Anesth Crit Care. 2023;

Chapter 11
Arterio-Venous Malformation

Abstract Arterio-venous malformations AVMs can be in any part of the body, the most common ones pulmonary, cerebral and hepatic.

Pulmonary AVMs are R → L shunts, small shunts are asymptomatic, larger ones can cause desaturation/ cyanosis and congestive heart failure.

Systemic AVMs are L → R shunts, small shunts are mostly asymptomatic, whereas larger ones cause congestive heart failure (high output failure).

Coronary AVMs cause a L → R shunt at coronary level and can cause coronary steal with myocardial ischaemia and heart failure. 50% of cases are asymptomatic.

Keywords Arterio-venous malformation · L → R shunt · Shunt lesion · R→L shunt · Vascular lesions · High output failure · Coronary abnormalities · Paediatric cardiac anaesthesia · Congenital cardiac lesion

Normal lung tissue:

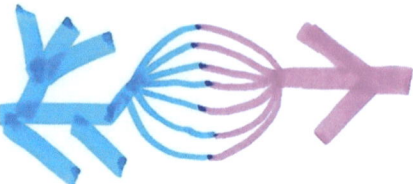

Fig. 11.1 Normal pulmonary capillaries: venous blood gets oxygenated in the capillaries around the alveoli

Pulmonary AVM:

Fig. 11.2 A pulmonary arterio-venous malformation does not participate in gas exchange and therefore deoxygenated blood creates a R → L shunt

Systemic capillaries:

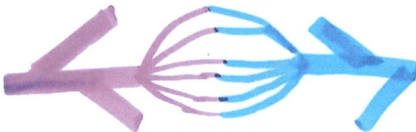

Fig. 11.3 Normal capillaries: body tissue gets oxygenated and deoxygenated blood returns to the lungs

Systemic AVM:

Fig. 11.4 An arterio-venous malformation will not participate in tissue oxygenation, causing tissue hypoxia and a higher venous saturation. In large lesions the lack of blood pressure generated by a capillary bed can cause high output failure

Arterio-venous malformations, also called arterio-venous fistulae, can be in the pulmonary or systemic circulation. Mild cases can be asymptomatic, moderate cases cause clinical symptoms depending on their location.

11.1 Pulmonary Arterio-Venous Malformation

- connection between pulmonary arteries and veins with R → L shunt (see Figs. 11.1 and 11.2)
- most common cause: hereditary haemorrhagic telangiectasia
- Clinical:
  - Mild: asymptomatic, mild desaturations, epistaxis
  - Large: cyanosis, clubbing, polycythaemia, congestive heart failure
  - Chest x-ray: can be normal, larger malformations can show up as opacity
  - ECG: normal, changes only in late stages of congestive heart failure
  - Echo: bubble study: intravenous injection of agitated water → bubbles appearing in left atrium
  - CT/ angiography: gold standard of detection

Anaesthetic Management
- R → L shunt, as the malformation does not participate in gas exchange
- Avoid increase in pulmonary vascular resistance: FiO2 > 21%, mild hyperventilation
- Avoid decrease in systemic blood pressure: careful anaesthetic, fluids, vasopressors
- Can deteriorate to high output failure
- Nitric Oxide might increase saturations

Anaesthesia for Repaired Lesion
- Usually no problems
- Pulmonary AVMs can re-occur

11.2 Systemic Arterio-Venous Malformation

- common in brain (Vein of Galen) and liver, but can appear in any body part (see Figs. 11.3 and 11.4)
- gastrointestinal AVMs are associated with aortic stenosis
- Clinical
  - Mild: asymptomatic
  - Large: congestive heart failure, high output failure, might hear flow murmur over the malformation

- ECG: Left ventricular hypertrophy (also right ventricular hypertrophy possible)
- Chest x-ray: cardiomegaly
- Echo: ventricular hypertrophy, aortic stenosis?

Anaesthetic Management
- Large lesions: **high risk anaesthetic** due to high output failure
- Avoid decrease in systemic blood pressure: fluids, vasopressors
- Inotropes usually not helpful

Anaesthesia for Repaired Lesions
- usually no problems

11.3 Coronary Arterio-Venous Malformation

- Rare, M:F = 2:1
- Coronary arterio-venous malformations are aberrant connections between a coronary artery and a major blood vessel or cardiac chamber
- In children 50% of coronary abnormalities are arterio-venous fistulae
- 90% congenital, 10% acquired (cardiac trauma, coronary surgery or cardiac disease like infarction or myocarditis)
- Right coronary artery: about 50%, left anterior descending artery: 35–40%
- 90% to venous circulation: most commonly to pulmonary arteries and right ventricle. Other sites: right or left atrium, coronary sinus, left ventricle, superior vena cava
- Pathophysiology: L $\rightarrow$ R shunt at coronary level, coronary steal, myocardial ischaemia, congestive heart failure, pulmonary hypertension
- Clinical:
    - 50% asymptomatic
    - Exertional dyspnoea
    - Chest pain, angina
    - ECG: normal or ST changes
    - Chest x-ray: normal, in congestive heart failure: cardiomegaly
    - Echo: coronary anatomy, wall motion abnormality, contractility/function
    - Angiography: detailed coronary anatomy
- Can be high risk anaesthesia, depending on severity

Anaesthetic Management
- Careful, haemodynamically stable anaesthesia
- Maintain preload
- Avoid tachycardia
- Keep sympathetic activity and myocardial oxygen demand low

Anaesthesia for Repaired Lesion
- Might have persistently distorted coronary arteries
- Ventricular impairment or failure can persist

Additional Reading

1. Hashimoto T, Young WL. Anesthesia-related considerations for cerebral arteriovenous malformations. FOC. 2001;11(5):1–6.
2. Teig MK. Anesthetic management of patients undergoing intravascular treatment of cerebral aneurysms and arteriovenous malformations. Anesthesiol Clin. 2021;39(1):151–62.
3. Majumdar S, McWilliams JP. Approach to pulmonary arteriovenous malformations: a comprehensive update. J Clin Med. 2020;9(6):1927.
4. Lee BB, Lardeo J, Neville R. Arterio-venous malformation: how much do we know? Phlebology. 2009;24(5):193–200.
5. Shovlin CL, Condliffe R, Donaldson JW, Kiely DG, Wort SJ. British Thoracic Society clinical statement on pulmonary arteriovenous malformations. Thorax. 2017;72(12):1154–63.
6. Hrishi A, Lionel K. Periprocedural management of vein of galen aneurysmal malformation patients: an 11-year experience. Anesth Essays Res. 2017;11(3):630–5.
7. Lakshmi BK, Dsouza S, Kulkarni A, Kamble J, Garasia M. Pulmonary arterio venous malformations – what the anesthesiologist must know. J Anaesthesiol Clin Pharmacol. 2019;35(2):271–3.

Chapter 12
Atrio-Ventricular Septal Defect, Complete

Abstract A complete atrio-ventricular defect consists of an atrial septal defect, a ventricular septal defect and a common atrio-ventricular valve. This results in a L → R shunt, both atrially and ventricularly. The common valve can be stenotic, regurgitant or both. Symptoms depend on the details of the lesions.

In some cases additional defects are associated with this, such as tetralogy of Fallot's or double outlet right ventricle. This creates a mixing lesion, which adds to the complexity of the anaesthetic management.

Keywords Atrio-ventricular septal defect · Atrial septal defect, ventricular septal defect · Valvular lesion, obstructive lesion, regurgitant lesion · L → R shunt, mixing lesion, cyanosis · Arrhythmia · Paediatric cardiac anaesthesia · Congenital cardiac lesion

Complete atrio-ventricular defects make up about 2% of congenital heart disease. It is also known as atrio-ventricular canal or endocardial cushion defect. Depending on the size of the septal defects, pressures and additional lesions this can be either a mixing lesion with systemic cyanosis or noncyanotic with a L → R shunt. The L → R shunt causes an increased pulmonary blood flow, leading to pulmonary hypertension, arrhythmias and congestive heart failure.

- 30% of defects occur with Trisomy 21
- Complete defect:
  - Atrial septal defect (ostium primum)
  - Ventricular septal defect (inlet)
  - Common atrio-ventricular valve
  - Balanced (Fig. 12.1): valve committed to both ventricles equally developed in size
  - Unbalanced: valve more committed to one ventricle with hypoplasia of the other ventricle

- Additional lesions
  - Double outlet right ventricle DORV ~6% (see Fig. 12.2)
  - Tetralogy of Fallot's ("canal tet") ~6% (see Fig. 12.3)
  - Transposition of the Great Arteries ~3%
- Clinical:
  - Congestive heart failure within 1–2 months of life
  - Tachypnoeic, poor feeding, failure to thrive
  - Heart murmur

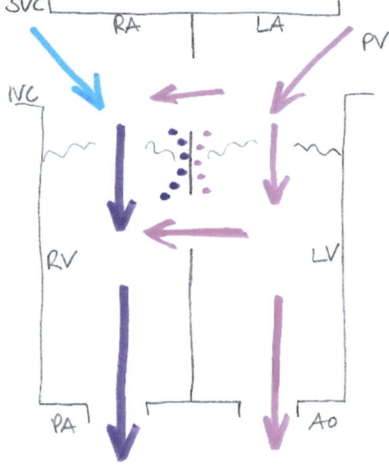

Fig. 12.1 Atrio-ventricular septal defect AVSD with L → R shunt. The dotted lines represent the possibility of valvular regurgitation or stenosis

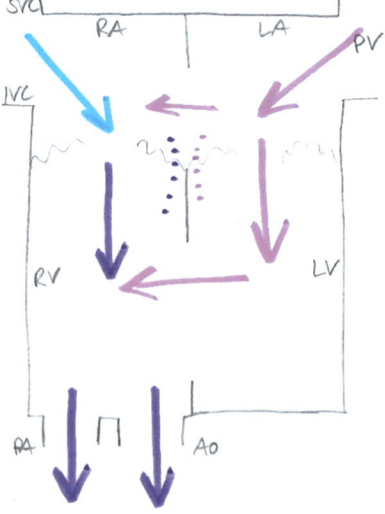

Fig. 12.2 AVSD with double outlet right ventricle DORV. The dotted lines represent the possibility of valvular regurgitation or stenosis. AVSD with double outlet right ventricle: L → R shunt, mixing lesion

Fig. 12.3 AVSD with Tetralogy of Fallot's. The dotted lines represent the possibility of valvular regurgitation or stenosis. AVSD with Tetralogy: in severely limited pulmonary blood flow might be duct-dependent, Mixing lesion

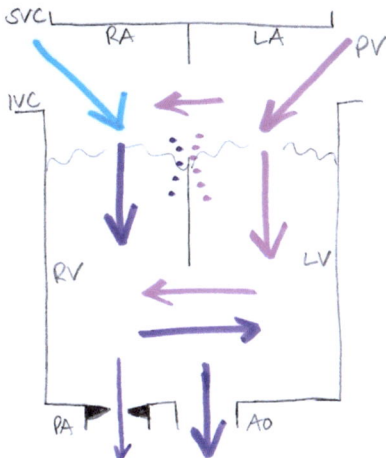

- ECG: superior QRS axis, right ventricular hypertrophy, right bundle branch block RBBB, atrioventricular blocks
- Chest x-ray: cardiomegaly, plethoric lungs
- ECHO: lesion in detail, size of septal defects, shunt volume, valve pathology, size and function of ventricles, etc.

Anaesthetic Management

- L → R shunt:
  - Avoid decrease in pulmonary vascular resistance: FiO2 21%, mild hypoventilation
  - Avoid increase in systemic vascular resistance: good anaesthesia and analgesia
- Atrio-ventricular valve can be stenotic, regurgitant or both → adjust heart rate accordingly
  - Stenosis: avoid tachycardia
  - Regurgitation: avoid bradycardia
- mixing lesion with systemic cyanosis:
  - cyanosis causes secondary erythrocytosis (Chap. 4):
    - coagulation disorder, making them prone to both thrombosis and bleeding
    - risk of acute hyperviscosity syndrome when dehydrated
    - keep well hydrated, minimise fasting times, start IV fluids preoperatively if necessary
  - balance pulmonary and systemic pressures for optimal blood flow and oxygenation.
- Inotropes might be needed for impaired function (unbalanced ventricles)

Anaesthesia for Repaired Lesion
- Unbalanced/DORV: two-ventricle repair not possible: single ventricle pathway with a 3-stage-repair:
  - pulmonary artery band (Chap. 41)
  - Glenn shunt (Chap. 38)
  - Total cavo-pulmonary connection TCPC (Chap. 44)
- Balanced: full repair of septal defects and valves. Additional repairs if needed: Tetralogy of Fallot's repair (Chap. 28) or switch (Chap. 43)
- Complications:
  - residual valve stenosis or regurgitation
  - residual shunt
  - arrhythmias can persist, might have pacemaker
- Check Echo for valve issues, ventricular function, residual shunts

Additional Reading

1. Mir AH, Ali Z, Dar BA, Naqash IA, Bashir S. Anesthetic management of a child with complete atrioventricular septal defect and single ventricle posted for noncardiac surgery. Anesth Essays Res. 2016;10(3):674–6.
2. Chauhan S. Atrioventricular septal defects. Ann Card Anaesth. 2018;21(1):1–3.
3. Taqatqa AS, Vettukattil JJ. Atrioventricular septal defects: pathology, imaging, and treatment options. Curr Cardiol Rep. 2021;23(8):93.
4. Xie O, Brizard CP, d'Udekem Y, Galati JC, Kelly A, Yong MS, et al. Outcomes of repair of complete atrioventricular septal defect in the current era. Eur J Cardiothorac Surg. 2014;45(4):610–7.
5. Ross FJ, Nasr VG, Joffe D, Latham GJ. Perioperative and anesthetic considerations in atrioventricular septal defect. Semin Cardiothorac Vasc Anesth. 2017;21(3):221–8.

Chapter 13
Atrio-Ventricular Septal Defect, Partial

Abstract Partial atrio-ventricular septal defects have an atrial septal defect and either a common atrioventricular valve or two valves. The atrioventricular valves can be stenotic or regurgitant and the atrial septal defect has a L → R shunt across it. These lesions can be asymptomatic, depending on the size of the shunt and the functionality of the valves. Late stage complications are pulmonary hypertension due to the L → R shunt as well as heart failure due to valve problems. Anaesthetic management needs to be adjusted to the individual lesion.

Keywords Atrio-ventricular septal defect, valvular lesion, arrhythmia · Atrial septal defect, shunt lesion · Obstructive lesion, regurgitant lesion · Paediatric cardiac anaesthesia · Congenital cardiac lesion

Partial atrio-ventricular septal defects make up about 1–2% of congenital heart disease. The L → R shunt causes an increased pulmonary blood flow, which can lead to pulmonary hypertension, arrhythmias and right heart failure.

- Does not have a ventricular septal component (no VSD, see Fig. 13.1)
- Atrial septal defect with either one common valve or two valves
- Symptoms depend on severity of lesion: shunt size and valve pathology
- Clinical:
  - Mostly asymptomatic
  - Heart murmur
  - ECG: superior QRS axis, right ventricular hypertrophy, right bundle branch block, etc.
  - Chest x-ray: might have cardiomegaly, plethoric lungs
  - Echo: details of defect, shunt size and flow, valve anatomy: regurgitation, stenosis with grading and severity
  - Large defects or late diagnosis: pulmonary hypertension and arrhythmias, cardiac dysfunction and failure

© The Author(s), under exclusive license to Springer Nature Switzerland AG 2025
J. Scheffczik, *Paediatric Cardiac Anaesthesia*,
https://doi.org/10.1007/978-3-031-90330-4_13

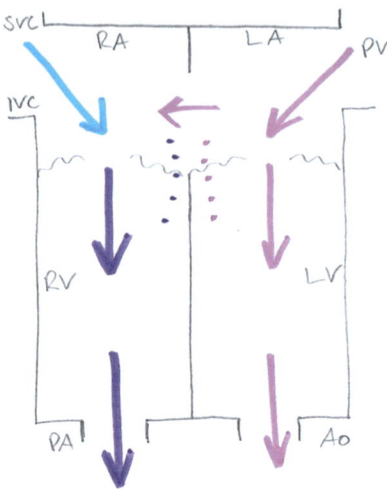

Fig. 13.1 Partial atrio-ventricular septal defect: no ventricular component. The dotted lines represent the possibility of valvular regurgitation or stenosis

Anaesthetic Management
- L → R shunt: avoid decrease of pulmonary vascular resistance: FiO2 21%, mild hypoventilation
- If regurgitant valves: avoid bradycardia
- If stenotic valves: avoid tachycardia
- Maintain preload

Anaesthesia for Repaired Lesion
- Usually good outcomes
- Check Echo for function, residual shunt or residual valve issues
- Arrhythmias?
- Pulmonary hypertension can persist

Additional Reading

1. Taqatqa AS, Vettukattil JJ. Atrioventricular septal defects: pathology, imaging, and treatment options. Curr Cardiol Rep. 2021;23(8):93.
2. Minich LL, Atz AM, Colan SD, Sleeper LA, Mital S, Jaggers J, et al. Partial and transitional atrioventricular septal defect outcomes. Ann Thorac Surg. 2010;89(2):530–6.

Chapter 14
Congenitally Corrected Transposition of the Great Arteries

Abstract Congenitally corrected transposition of the great arteries may stay asymptomatic and undetected. Most ccTGAs have associated lesions like a ventricular septal defect or pulmonary stenosis, which causes symptoms and complications. Another common problem are arrhythmias and varying degrees of heart block, due to conduction anomalies. Anaesthetic management needs to be adjusted to the individual lesion, dealing with a L → R shunt or a mixing lesion. In the late stages heart failure is common and arrhythmias can occur at any time.

Keywords Congenitally corrected TGA · ventricular septal defect · pulmonary stenosis, conduction abnormalities, arrhythmia · Paediatric cardiac anaesthesia · Congenital cardiac lesion

Easy, straightforward anaesthesia

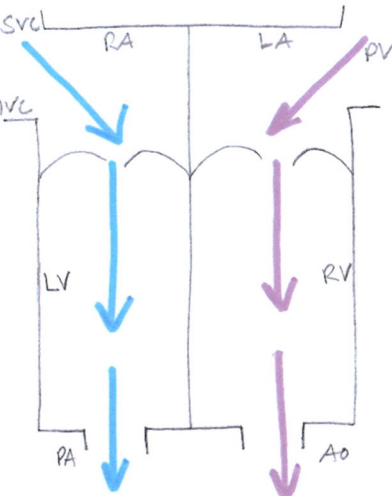

Fig. 14.1 Congenitally corrected transposition of the great arteries ccTGA

© The Author(s), under exclusive license to Springer Nature Switzerland AG 2025
J. Scheffczik, *Paediatric Cardiac Anaesthesia*,
https://doi.org/10.1007/978-3-031-90330-4_14

Congenitally corrected transposition of the great arteries is a rare lesion, <1% of congenital heart disease. It is also referred to as L-TGA (Levo-transposition). Associated cardiac lesions can complicate anaesthesia and surgical options for repair.

- If no associated anomalies: asymptomatic, may stay undetected (Fig. 14.1)
- Associated cardiac anomalies:
    - 70–80% Ventricular Septal Defect (Fig. 14.2)
    - 40–50% Pulmonary stenosis (Fig. 14.3)
    - 30% Tricuspid regurgitation TR
    - varying degrees of arrhythmia, like atrio-ventricular block and supraventricular tachycardia SVT

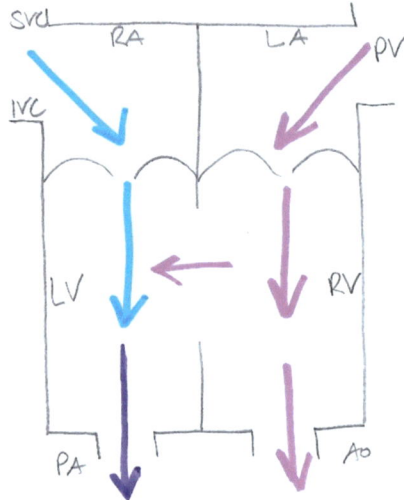

Fig. 14.2 ccTGA with ventricular septal defect VSD

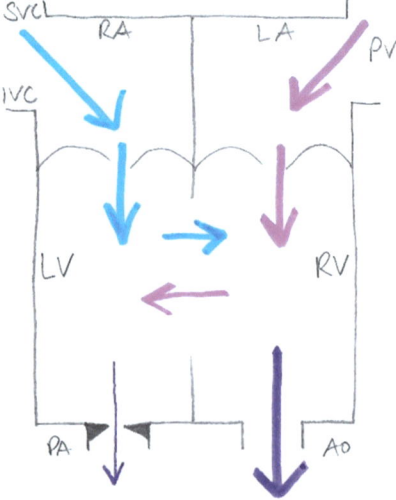

Fig. 14.3 ccTGA with ventricular septal defect VSD and pulmonary stenosis

14 Congenitally Corrected Transposition of the Great Arteries

- Clinical:
  - Dependent on associated lesion:
    - Cyanosis (VSD and PS).
    - Congestive heart failure (VSD)
  - Murmur, dependent on lesion
  - ECG: might have AV block, hypertrophy of atria and ventricles
  - Chest x-ray: cardiomegaly with VSD
  - Echo: shows lesion and gradients
  - Tricuspid regurgitation TR develops in 30% of patients
  - AV block and complete heart block in 30% of patients

Anaesthetic Management
- dependent on lesion and gradient
- CCTGA and VSD (Fig. 14.2): treat as VSD: L → R shunt:
  - Avoid decrease in pulmonary vascular resistance: FiO2 21%, mild hypoventilation
  - Avoid increase in systemic blood pressure: good anaesthesia and analgesia
- CCTGA, VSD, PS (Fig. 14.3):
  - Pulmonary stenosis limits pulmonary blood flow ➔ dependent on pressure gradient and VSD might be L → R shunt or R → L shunt with mixing/cyanosis
  - cyanosis causes secondary erythrocytosis (Chap. 4):
    - coagulation disorder, making them prone to both thrombosis and bleeding
    - risk of acute hyperviscosity syndrome when dehydrated
    - keep well hydrated, minimise fasting times, start IV fluids preoperatively if necessary
  - Decrease pulmonary vascular resistance as needed: FiO2 > 21%, mild hyperventilation
  - Avoid tachycardia
- Prepare for arrhythmia control: medication and pacing available
- Might need inotropes for heart failure
- Tricuspid regurgitation: avoid bradycardia

Anaesthesia for Repaired Lesion
- Different surgical options dependent on details of lesion
- Asymptomatic lesions might stay unrepaired until symptomatic
- Palliative procedures:
  - Pulmonary artery band (Chap. 41)
  - BT shunt (Chap. 36) for severe pulmonary stenosis or atresia
- No VSD or left ventricular outflow tract obstruction: left ventricle needs training to provide systemic pressures: pulmonary artery band (Chap. 41) prior to full repair (double switch, Chap. 43)

- No tricuspid regurgitation or right ventricular dysfunction: right ventricle stays systemic ventricle, VSD closure, with or without pulmonary stenosis relief
  - Residual VSD?
  - Might need inotropes for ventricular function impairment
  - Residual pulmonary stenosis: avoid tachycardia, decrease pulmonary blood pressure if needed with FiO2 > 21% and mild hyperventilation
- Tricuspid regurgitation and/or right ventricular dysfunction: repair makes left ventricle the systemic ventricle: Double switch (Chap. 43)
- Complex lesions → single ventricle pathway: Glenn shunt (Chap. 38), then total cavo-pulmonary connection TCPC (Chap. 44)

Additional Reading

1. Pahuja HD, Gadkari CP, Wakode NG, Bhure AR. Anaesthetic management of a case of congenitally corrected transposition of great arteries for non cardiac surgery: a case report. IJCA. 2021;8(2):341–4.
2. Chen J, Tan SH, Chee SWL, Kothandan H. Anaesthetic management of a patient with complex, cyanotic congenitally corrected transposition of great arteries for electrophysiological study and thermoablation. BMJ Case Rep. 2022;15(4):e247265.
3. Naoum EE, Ortoleva JP, Militana RM, Soffer MD, Yeh DD. Anesthesia for cesarean delivery in a patient with congenitally corrected transposition of the great arteries: a case report. Ann Card Anaesth. 2023;26(4):446–50.
4. Arendt KW, Connolly HM, Warnes CA, Watson WJ, Hebl JR, Craigo PA. Anesthetic management of parturients with congenitally corrected transposition of the great arteries: three cases and a review of the literature. Anesth Analg. 2008;107(6):1973–7.
5. Kumar TKS. Congenitally corrected transposition of the great arteries. J Thorac Dis. 2020;12(3):1213–8.
6. Kutty S, Danford DA, Diller GP, Tutarel O. Contemporary management and outcomes in congenitally corrected transposition of the great arteries. Heart. 2018;104(14):1148–55.

Chapter 15
Coarctation of the Aorta/Hypoplastic Aortic Arch

Abstract Coarctation is a stenosis of the descending aorta at the level of the duct insertion. This causes hypertension in the upper limbs and head, and hypotension in the abdomen and lower limbs. If undetected this can lead to left ventricular failure, pulmonary hypertension and hypertensive encephalopathy. Anaesthetic management of the stenosis is to avoid tachycardia and keeping an adequate blood pressure for renal perfusion. Inotropes might be needed, depending on ventricular function. Associated cardiac lesions will need additional considerations.

Keywords Vascular lesion · Aorta · Aortic arch abnormalities · Aortic arch · Obstructive lesion · Coarctation · Aortic arch hypoplasia · Paediatric cardiac anaesthesia · Congenital cardiac lesion

15.1 Coarctation of the Aorta

Coarctation accounts for about 4–8% of congenital heart disease, more common in males than females. Clinically it is diagnosed with the combination of upper limb hypertension and weak or absent femoral pulses.

- Stenosis of the descending aorta at the level of the duct insertion (Fig. 15.1)
- Might have associated cardiac anomalies:
  - bicuspid aortic valve BAV: about 85%
  - hypoplastic aortic arch: about 80%
  - Taussig-Bing anomaly
- Difficult to detect in presence of persisting ductus arteriosus

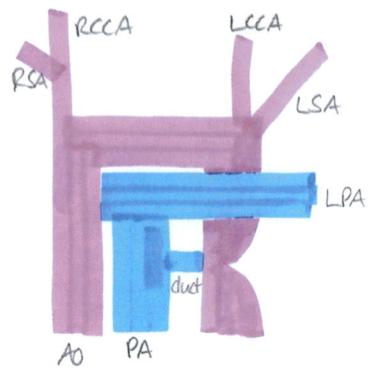

Fig. 15.1 Coarctation: stenosis of the descending aorta Ao juxtaposed to the duct insertion. *Ao* aorta, *PA* pulmonary artery, *RSA* right subclavian artery, *RCCA* right common carotid artery, *LCCA* left common carotid artery, *LSA* left subclavian artery

- Pseudocoarctation:
  - very rare, associated with cardiac conditions like bicuspid aortic valve, cervical aortic arch, etc. and genetic abnormalities
  - elongated aortic arch with kinking at the level of the ligamentum arteriosum
  - Usually presents with blood pressure differences: hypertension in the upper body and low blood pressure in the lower limbs
  - Complications: aortic aneurysm
- Clinical:
  - Asymptomatic infants:
    - Upper limb hypertension
    - Might have murmur (bicuspid aortic valve)
    - ECG: normal or left ventricular hypertrophy
    - Chest x-ray: normal or mildly enlarged heart, "fig. 3" sign in aortic arch outline
    - Echo: Coarctation, associated lesions, contractility
  - Symptomatic infants:
    - Congestive heart failure
    - Difficulty feeding and failure to thrive
    - Renal failure due to low blood pressure
    - ECG: right ventricular hypertrophy, right bundle branch block RBBB
    - Chest x-ray: "fig. 3" sign in aortic arch outline, cardiomegaly, pulmonary congestion
    - Echo: Coarctation, associated lesions, contractility
    - late symptoms: left ventricular failure, intracranial haemorrhage, pulmonary hypertension, hypertensive encephalopathy

Anaesthetic Management
- Keep well filled, maintain preload
- Head up position might be beneficial

- Blood pressure measurement on upper and lower limbs
- Keep heart rate low, avoid tachycardia
- Might need inotropes to improve function and forward flow
- Anaesthetic management for associated lesions

Anaesthesia for Repaired Lesion
- Usually good long-term results
- Arterial hypertension might persist
- Check Echo for repair as well as associated lesions and their repair:
  - Re-coarctation? (about 10–20% of cases)
  - Development of aneurysms or pseudoaneurysms can occur
  - Coronary artery disease is more common in repaired coarctation patients
- Surgical techniques using the left subclavian artery (subclavian flap) for the repair might result in a reduced pulse on the left arm due to perfusion via collaterals
  - might lead to arm claudication: careful positioning of the arm during anaesthesia
  - arterial blood pressure measurement in the left wrist might not be accurate

15.2 Hypoplastic Aortic Arch

Aortic arch hypoplasia is rare as an isolated lesion. It is commonly associated with lesion of aortic arch- adjacent structures, such as aortic valve dysplasia or atresia, hypoplastic left heart syndrome, coarctation, etc.

- Hypoplasia can occur at any point of the aortic arch: ascending, transverse, descending
- Can be a short or long segment
- Can involve/impair ostia of arteries arising from the aorta, such as coronaries or upper body arteries (subclavian steal syndrome)
- Symptoms depending on position and severity of hypoplasia: hypertension and dilatation before obstruction, hypotension distal to the hypoplastic part
- Clinical:
  - Hypotension distal to hypoplasia
  - Symptoms of associated lesions, such as cyanosis, cardiac dilatation and dysfunction, etc.
  - In subclavian steal syndrome: exertional headache, asymmetrical pulses in both arms
  - Late stages: congestive heart failure, pulmonary hypertension
  - Might have flow murmur or murmur of associated lesions
  - ECG: nothing specific
  - Chest x-ray: aortic arch hypoplasia can rarely be seen

- Echo: details of position and length of lesion, blood flow and gradient of obstruction, pre-stenotic dilation, ostia of branching arteries, coronary ostia and blood flow, associated lesions

Anaesthetic Management
- avoid tachycardia: good anaesthesia and analgesia
- maintain preload
- Blood pressure measurement on upper and lower limbs, depending on severity of obstruction
- Keep coronary perfusion pressure
- Inotropes if needed to improve function and forward flow
- Anaesthetic management for associated lesions

Anaesthesia for Repaired Lesion
- Surgical repair depends on position and length of hypoplasia:
  - End to end anastomosis in short segment hypoplasia
  - Patch to enlarge aorta
  - Homograft or bioprosthetic patches for long segment hypoplasia
- Repair and outcomes depend on severity and associated lesions
- Long suture lines and anastomosis can calcify, narrow and become stenotic over time
- Check Echo:
  - Aortic arch size and blood flow
  - Residual or new obstructions?
  - Cardiac function/ contractility: if lesion discovered late, cardiac function might not recover
  - Coronary blood flow
  - Blood flow to any arteries involved in initial hypoplastic segment
  - Associated lesions?

Additional Reading

1. Meidell Blylod V, Rinnström D, Pennlert J, Ostenfeld E, Dellborg M, Sörensson P, et al. Interventions in adults with repaired coarctation of the aorta. JAHA. 2022;11(14):e023954.
2. Fox EB, Latham GJ, Ross FJ, Joffe D. Perioperative and anesthetic Management of coarctation of the aorta. Semin Cardiothorac Vasc Anesth. 2019;23(2):212–24.

Chapter 16
Cor triatriatum

Abstract Cor triatriatum is a rare condition where the left atrium is septated by a fibro-muscular membrane, causing obstruction. Symptoms depend on degree of obstruction, from asymptomatic in minor obstruction to low cardiac output failure and pulmonary congestion with pulmonary hypertension in severe obstruction. Some cases have an associated atrial septal defect, causing a L→R shunt. Anaesthetic management needs to be adjusted to the individual lesion, avoiding tachycardia and avoiding an increase in systemic blood pressure for promoting forward flow.

Keywords Cor triatriatum · Obstructive lesion · L→R shunt · Shunt lesion · Paediatric cardiac anaesthesia · Congenital cardiac lesion

Cor triatriatum is a very rare lesion, with a fibro-muscular membrane septating the left atrium, causing obstruction (Fig. 16.1). Symptoms depend on the gradient of the obstruction and the presence of an atrial septal defect.

- Haemodynamically like mitral stenosis
    - Pulmonary venous and arterial congestion and hypertension
- Might have atrial septal defect ASD (Figs. 16.2 and 16.3): L→R shunt with increased pulmonary blood flow, aggravating pulmonary hypertension
- Clinically:
    - Mostly asymptomatic unless severe obstruction or large ASD
    - Dyspnoea, tachypnoea
    - Low cardiac output syndrome, congestive heart failure
    - Weak peripheral pulses
    - Pulmonary oedema, pulmonary hypertension
    - ECG: might have atrial and ventricular hypertrophy
    - Chest x-ray: pulmonary congestion and oedema
    - Echo: membrane and its relationship to atrial septal defect, associated defects

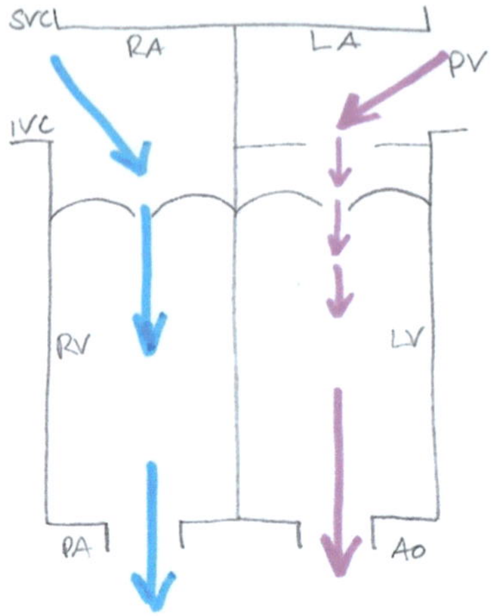

Fig. 16.1 Cor triatriatum without atrial septal defect ASD: treat like mitral stenosis

Fig. 16.2 Cor triatriatum with atrial septal defect ASD above the membrane: treat like L→R shunt with mitral stenosis

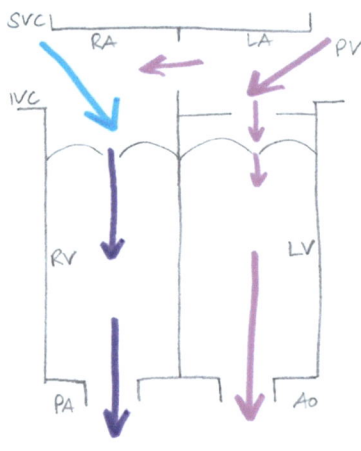

Fig. 16.3 Cor triatriatum with atrial septal defect ASD below the membrane: treat like L→R shunt with mitral stenosis

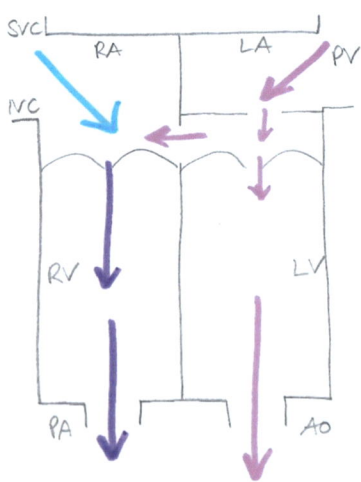

Anaesthetic Management
- No ASD: Treat like mitral stenosis/ obstruction:
  - Avoid tachycardia
  - Maintain systemic blood pressure for good coronary and peripheral perfusion
- With ASD: treat like L→R shunt with mitral stenosis:
  - Avoid tachycardia
  - Avoid decrease in pulmonary vascular resistance PVR: FiO2 21%, mild hypoventilation
  - Avoid increase in systemic vascular resistance SVR: good anaesthetic
- Maintain preload
- Might need intropes, depending on heart function

Anaesthesia for Repaired Lesion
- No pre-cautions necessary
- Arrhythmias possible
- Might have persistent pulmonary hypertension

Additional Reading

1. Sprung J, Scavonetto F, Yeoh T, Welch T, Weingarten T. Anesthesia and cor triatriatum. Ann Card Anaesth. 2014;17(2):111.
2. Jha AK, Makhija N. Cor Triatriatum: a review. Semin Cardiothorac Vasc Anesth. 2017 Jun;21(2):178–85.
3. Tewari P. Cor-triatriatum: when to worry? Ann Card Anaesth. 2014;17(2):116–7.

Chapter 17
Double Inlet Left Ventricle

Abstract Double inlet left ventricle (DILV) is a complex cyanotic lesion with several pathophysiological components. Both atrioventricular valves empty into one ventricle, with the other ventricle rudimentary and connected by a bulbo-ventricular foramen. Most cases are associated with additional cardiac defects, most commonly with a transposition of the great arteries or pulmonary stenosis or atresia.

Anaesthetic management depends on the individual lesions and their severity. This is a mixing lesion with varying degrees and locations of obstruction, resulting in either a decreased systemic or decreased pulmonary blood flow.

Keywords DILV · Mixing lesion complex cyanotic lesion · Shunt lesion · Single ventricle, cyanosis · Paediatric cardiac anaesthesia · Congenital cardiac lesion

Double inlet left ventricle is a rare but complex cyanotic lesion, at less than 1% of congenital heart disease. Symptoms vary with number and severity of associated lesions, which influences pulmonary blood flow. Anaesthetic management mirrors the complexity of the lesion.

- Both atrio-ventricular valves empty into one chamber/ ventricle
- One ventricle is rudimentary and connected by bulbo-ventricular foramen BVF ("VSD") → in 80% main ventricle is the left ventricle
- Associated lesions:
  - 85% Transposition of the great arteries TGA or congenitally corrected transposition of the great arteries
  - 50% Pulmonary stenosis or atresia
  - May have Coarctation or interrupted arch
    - Most common: DILV with TGA
- Bulbo-ventricular foramen becomes obstructive over time

- Clinically:
  - Mixing lesion with cyanosis, oxygen saturations 75–85%
  - Symptoms depends on pulmonary blood flow PBF
  - High PBF: similar to transposition with large ventricular septal defect
  - Low PBF: similar to Tetralogy of Fallot's
  - Congestive heart failure, failure to thrive
  - Pulmonary congestion: dyspnoea, tachypnoea
  - ECG: atrio-ventricular blocks, arrhythmia, ventricular hypertrophy
  - Chest x-ray:
    - High PBF: cardiomegaly, pulmonary congestion, plethoric lungs
    - Low PBF: normal
  - Echo: ventricular arrangement and contractility/function, size and gradient of bulbo-ventricular foramen, great arteries arrangement, additional lesions like pulmonary stenosis with gradient
- **High risk anaesthetic**

Anaesthetic Management

- Chronic cyanosis causes secondary erythrocytosis (Chap. 4):
  - coagulation disorder, making them prone to both thrombosis and bleeding
  - risk of acute hyperviscosity syndrome when dehydrated
  - keep well hydrated, minimise fasting times, start IV fluids preoperatively if necessary
- might need inotropes for decreased contractility
- caveat arrhythmias, might need medication or pacing
- anaesthetic management dependent on details of lesion: see Figs. 17.1, 17.2, 17.3, and 17.4:

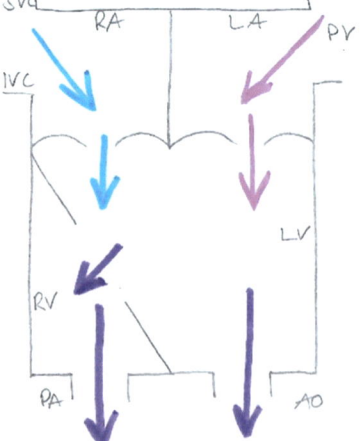

Fig. 17.1 Double inlet left ventricle DILV, unobstructed bulbo-ventricular foramen

17 Double Inlet Left Ventricle

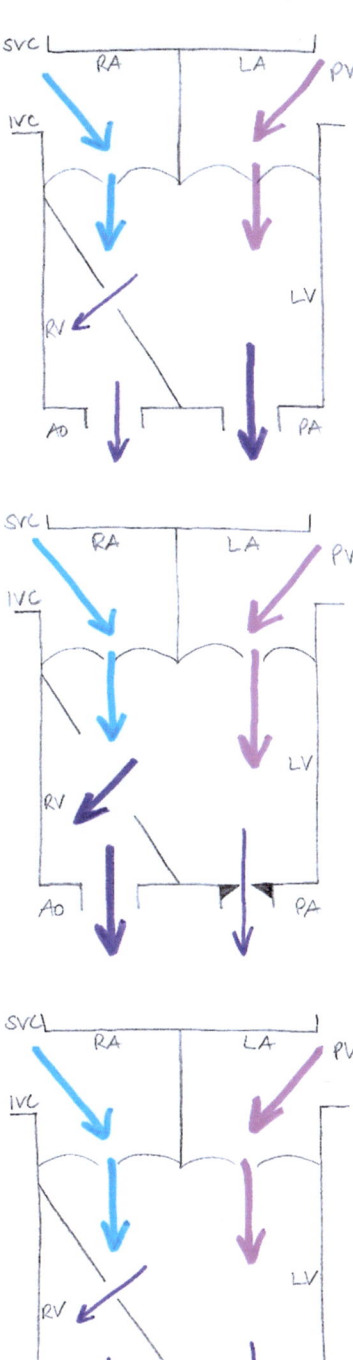

Fig. 17.2 Double inlet ventricle DILV, transposition of the great arteries TGA, obstructed bulbo-ventricular foramen

Fig. 17.3 Double inlet left ventricle DILV with transposition of the great arteries TGA, pulmonary stenosis, unobstructed bulbo-ventricular foramen

Fig. 17.4 Double inlet left ventricle DILV with transposition of the great arteries TGA, pulmonary stenosis, obstructed bulbo-ventricular foramen

- No pulmonary stenosis, unobstructed bulbo-ventricular foramen (Fig. 17.1)
    - mixing lesion, oxygen saturations 75–85%
    - BVF: L→R shunt: avoid decrease in pulmonary pressures, avoid increase in systemic blood pressure
- DILV with transposition, no pulmonary stenosis, obstructed BVF: decreased systemic blood flow (Fig. 17.2)
    - Mixing lesion, oxygen saturations 75–85%
    - avoid decrease in pulmonary vascular resistance
    - avoid increase in systemic vascular resistance
    - avoid tachycardia
- DILV with transposition, pulmonary stenosis, unobstructed BVF: decreased pulmonary blood flow (Fig. 17.3)
    - Mixing lesion, oxygen saturations 75–85%
    - Pulmonary stenosis: decrease pulmonary vascular resistance: FiO2 >21%, mild hyperventilation
- DILV with transposition, pulmonary stenosis, obstructed BVF: biventricular outflow tract obstruction (Fig. 17.4) → **high risk anaesthetic**
    - Mixing lesion, oxygen saturations 75–85%
    - promote forward flow by decreasing pulmonary blood pressure: FiO2 >21%, mild hyperventilation
    - maintain systemic blood pressure with haemodynamically stable anaesthetic and fluids for coronary perfusion
    - might need inotropic support

Anaesthesia for Repaired Lesion
- Three stage repair to a single ventricle circulation
    - First stage: dependent on lesion:
    - No PS, unobstructed BVF: pulmonary artery banding to limit L→R shunt to prevent pulmonary hypertension (Chap. 41)
    - No PS, obstructed BVF: Damus-Kaye-Stansel (Chap. 37) with Blalock-Taussig shunt (Chap. 36)
    - PS, unobstructed BVF: Blalock-Taussig shunt (Chap. 36) or ductal stent (Chap. 41)
    - PS, obstructed BVF: enlargement of BVF and Blalock-Taussig shunt (Chap. 36)
    - Second stage: Glenn shunt (Chap. 38)
    - Third stage: Total cavo-pulmonary connection TCPC (Chap. 44)

Additional Reading

1. Nicolson SC, Steven JM, Diaz LK, Andropoulos DB. Anesthesia for the patient with a single ventricle. In: Andropoulos DB, Stayer S, Mossad EB, Miller-Hance WC, editors. Anesthesia for congenital heart disease [Internet]. 1st ed. Wiley; 2015 [cited 2024 Apr 27]. p. 567–97. https://onlinelibrary.wiley.com/doi/10.1002/9781118768341.ch25
2. Rao PS. Double-inlet left ventricle. Children. 2022 Aug 24;9(9):1274.
3. Greaney D, Honjo O, O'Leary JD. The single ventricle pathway in paediatrics for anaesthetists. BJA Educ. 2019 May;19(5):144–50.

Chapter 18
Double Outlet Right Ventricle

Abstract Double outlet right ventricle (DORV) is a complex cyanotic lesion. Both the pulmonary artery and the aorta arise from the right ventricle with a ventricular septal defect emptying the left ventricle. Symptoms and anaesthetic management depend on the position of the ventricular septal defect and any additional lesions and their severity. This is a mixing lesion with a L→R shunt and complications such as reduced pulmonary blood flow with varying degrees of cyanosis, congestive heart failure or pulmonary hypertension.

Keywords DORV · Complex cyanotic lesion · Mixing lesion, cyanosis · Shunt lesion · Single ventricle · Taussig-Bing anomaly · Paediatric cardiac anaesthesia · Congenital cardiac lesion

Double outlet right ventricle is a rare, but complex lesion, at less than 1% of congenital heart disease. Saturations depend on the individual lesion and blood flow. Anaesthetic management mirrors the complexity of the lesion.

- Pulmonary artery and Aorta arise from right ventricle (Fig. 18.1)
- Types depending on position of ventricular septal defect VSD:
  - Sub-aortic VSD (Fig. 18.2)
  - Sub-pulmonary VSD = Taussig-Bing Anomaly (Fig. 18.3)
  - Sub-aortic VSD and pulmonary stenosis (Fig. 18.4)
  - Doubly committed or remote VSD

Fig. 18.1 Double outlet right ventricle

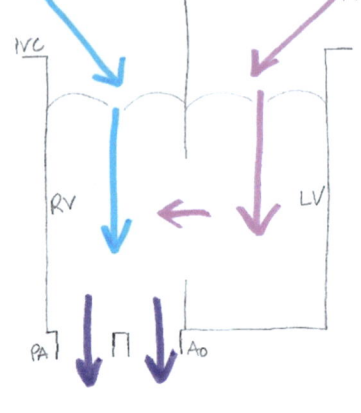

Fig. 18.2 Double outlet right ventricle with subaortic ventricular septal defect

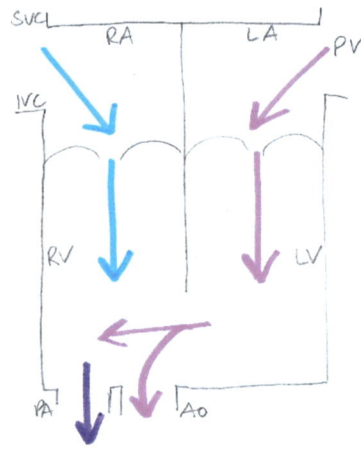

Fig. 18.3 Double outlet right ventricle with subpulmonary ventricular septal defect

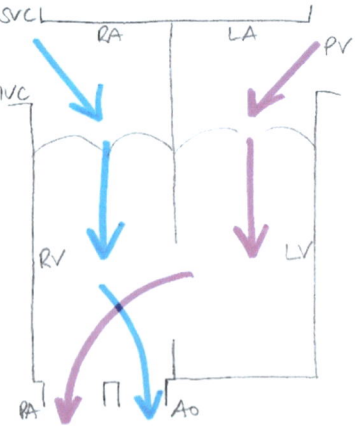

Fig. 18.4 Double outlet right ventricle with pulmonary stenosis: similar to Tetralogy of Fallot's

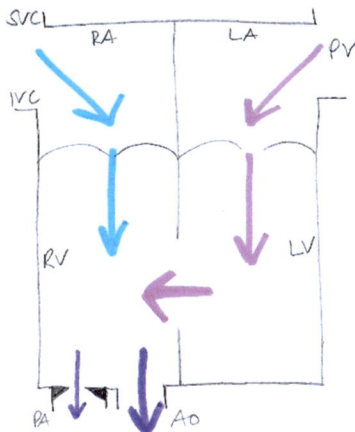

- Clinical: depending on size and position of ventricular septal defect:
    - Sub-aortic: mild cyanosis, pulmonary hypertension, congestive heart failure
    - Sub-pulmonary: severe cyanosis, congestive heart failure
    - Sub-aortic with pulmonary stenosis: cyanosis, reduced pulmonary blood flow
    - Doubly committed or remote: mild cyanosis
    - Murmur of ventricular septal defect and/ or pulmonary stenosis
    - ECG: can be normal
    - Chest x-ray: cardiomegaly, may have pulmonary congestion
    - Echo: size and location of ventricular septal defect (unrestricted?), pulmonary hypertension (Tricuspid regurgitation), volume load and dilatation, function/contractility, additional lesions?
- **High risk anaesthetic**

Anaesthetic Management
- Dependent on details of lesion
- Chronic cyanosis causes secondary erythrocytosis (Chap. 4):
    - coagulation disorder, making them prone to both thrombosis and bleeding
    - risk of acute hyperviscosity syndrome when dehydrated
    - keep well hydrated, minimise fasting times, start IV fluids preoperatively if necessary
- subaortic VSD: (Fig. 18.2) L→R shunt
    - avoid decrease in pulmonary vascular resistance: FiO2 21%, mild hypoventilation
    - avoid increase in systemic vascular resistance: good analgesia and anaesthesia
- sub-pulmonary VSD (Fig. 18.3): left ventricular blood flow gets directed to pulmonary artery, right ventricular blood flow is directed to aorta:
    - keep pulmonary and systemic vascular resistance at preoperative levels
    - change pressures as needed to balance both circulations

- Doubly committed/ remote VSD:
    - keep pulmonary and systemic vascular resistance at preoperative levels
    - change pressures as needed to balance both circulations
- Sub-aortic VSD with pulmonary stenosis: (Fig. 18.4): like Tetralogy of Fallot's:
    - decrease pulmonary vascular resistance PVR with FiO2 >21% and mild hyperventilation
    - avoid tachycardia
    - maintain ductal patency in critical stenosis: Prostaglandin infusion
    - if patent duct: L→R shunt:
        - avoid decrease in pulmonary vascular resistance: FiO2 21%, mild hypoventilation
        - avoid increase in systemic vascular resistance: good analgesia and anaesthesia

Anaesthesia for Repaired Lesions
- Usually attempt at full repair
- Sub-aortic/ doubly committed VSD: closure of ventricular septal defect to achieve VSD-Aortic tunnel
- Subaortic VSD with pulmonary stenosis: Blalock-Taussig shunt (Chap. 36), then VSD-Aortic tunnel and Rastelli (Chap. 43)
- Sub-pulmonary: atrial septal defect enlargement or septostomy, then VSD-Pulmonary artery tunnel and switch (Chap. 43)
- Multiple or remote ventricular septal defects: Pulmonary artery band (Chap. 41), then VSD-Aortic tunnel
- If hypoplastic ventricle: single ventricle pathway with a 3-stage repair: Blalock-Taussig shunt (Chap. 36) → Glenn shunt (Chap. 38) → Total cavo-pulmonary connection TCPC (Chap. 44)

Additional Reading

1. Kotwani M, Rayadurg V, Tendolkar B. Anaesthetic management in a patient of uncorrected double outlet right ventricle for emergency surgery. Indian J Anaesth. 2017;61(1):87.
2. Gu J, Cai Y, Liu B, Lv S. Anesthetic management for cesarean section in a patient with uncorrected double-outlet right ventricle. SpringerPlus. 2016 Dec;5(1):415.
3. Krishna R, Goneppanavar U. Anesthetic management of caesarean section in a patient with double outlet right ventricle. J Obstet Anaesth Crit Care. 2012;2(1):50.
4. Anesthesiology and Intensive Therapy Department, Medical Faculty, Brawijaya University/Dr. Saiful Anwar General Hospital, Malang, Indonesia Indonesia, Bhirowo BK, Vitraludyono R, Anesthesiology and Intensive Therapy Department, Medical Faculty, Brawijaya University/Dr. Saiful Anwar General Hospital, Malang, Indonesia. Management anesthesia of esophagostomy in a patient with a double outlet right ventricle. J Anaesth Pain. 2021 May 30;2(2):89–92.
5. Greaney D, Honjo O, O'Leary JD. The single ventricle pathway in paediatrics for anaesthetists. BJA Educ. 2019 May;19(5):144–50.

Chapter 19
Ebstein's Anomaly/Malformation

Abstract Ebstein's anomaly is a valvular lesion. The tricuspid valve is dysplastic and displaced downward into the right ventricle with variable degree of severity and symptoms. The right atrium is dilated and may cause atrial arrhythmias. The tricuspid valve might be stenotic or regurgitant or both. The size and function of the right ventricle depends on the amount of valve displacement and might be in failure.

Anaesthetic management is dependent on the individual lesion and can be challenging in severe cases.

Keywords Ebstein's anomaly Ebstein's malformation · Valvular lesion, valve lesion · Tricuspid valve · Heart failure · Arrhythmia · Paediatric cardiac anaesthesia · Congenital cardiac lesion

Ebstein's anomaly is a rare (< 1% of congenital heart disease) malformation of the tricuspid valve and the right ventricle. Symptoms and repair depend on the severity of the lesion.

- Posterior and septal leaflet of tricuspid valve are displaced downwards into right ventricle (Fig. 19.1)
- Degrees of severity → variety of symptoms:
  - Functional hypoplasia of right ventricle as part of it is atrialised: ventricular dysfunction
  - Right atrium dilated and hypertrophied → arrhythmias
  - Atrial septal defect or Patent foramen ovale: R → L shunt (Fig. 19.2) with varying degrees of cyanosis
  - Tricuspid valve: regurgitation or stenosis or both
  - Fibrosis of ventricular free walls → ventricular dysfunction
  - Conduction abnormalities: associated with Wolff-Parkinson-White (WPW) syndrome, supraventricular tachycardia SVT and other arrhythmias

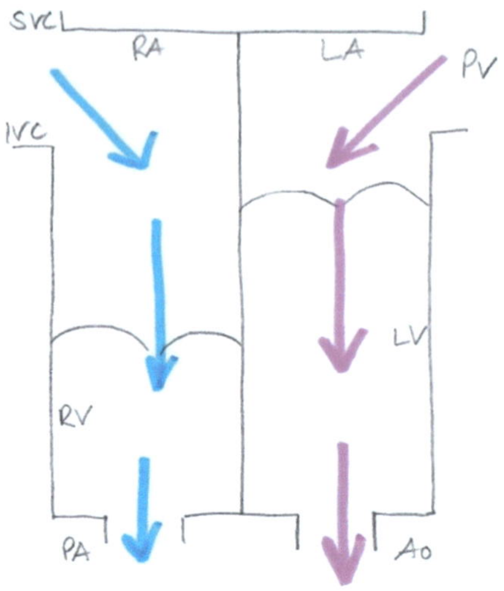

Fig. 19.1 Ebstein's anomaly: part of the tricuspid valve is displaced downwards into the right ventricle. The valve can be dysplastic, causing obstruction or regurgitation

Fig. 19.2 Ebstein's anomaly with atrial septal defect ASD or persistent foramen ovale PFO

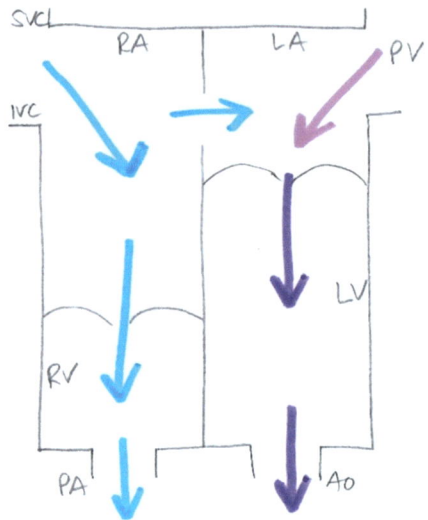

- Clinical: dependent on severity
    - Cyanosis, congestive heart failure
    - Right ventricular pressure < Pulmonary vascular resistance: no forward flow through pulmonary artery → duct dependent (Fig. 19.3)
    - Might have murmur of atrial septal defect or duct
    - ECG: right bundle branch block RBBB, right atrial hypertrophy, Wolff-Parkinson-White syndrome, supraventricular tachycardia SVT, atrio-ventricular block
    - Chest x-ray: cardiomegaly

Fig. 19.3 Severe Ebstein's, duct-dependent, as right ventricular pressure is lower than pulmonary pressure, there's no forward flow into pulmonary arteries from right ventricle. As pulmonary vascular resistance decreases in the first weeks of life, improvement is possible

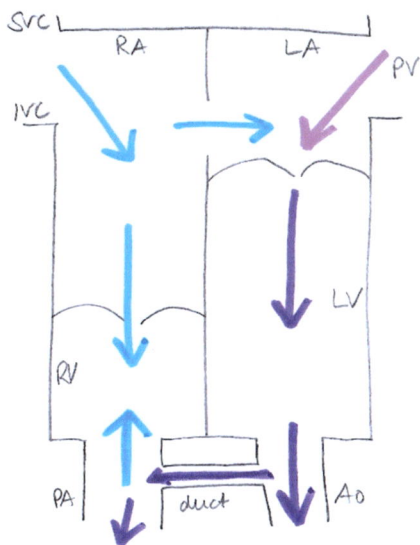

- Echo: severity of tricuspid valve displacement, contractility/ function of right ventricle, tricuspid valve anatomy and function (stenosis? regurgitation?), septal defects and shunt volume, additional lesions?

- **high risk anaesthesia for severe cases**

Anaesthetic Management
- promote forward flow to pulmonary artery: decrease pulmonary vascular resistance with FiO2 > 21% and mild hyperventilation
- avoid increase in systemic vascular resistance: good anaesthesia and analgesia
- if arrhythmias: attach defibrillator pads, have adenosine for SVT and pacing available
- if tricuspid regurgitation: avoid bradycardia
- if tricuspid stenosis: avoid tachycardia
- might need inotropes
- if duct-dependent:

    - Prostaglandin infusion
    - L → R shunt:

        - avoid decrease in pulmonary vascular resistance: FiO2 21% and mild hypoventilation
        - avoid increase in systemic vascular resistance: good analgesia and anaesthesia

- if R → L shunt: cyanosis causes secondary erythrocytosis (Chap. 4):
  - coagulation disorder, making them prone to both thrombosis and bleeding
  - risk of acute hyperviscosity syndrome when dehydrated
  - keep well hydrated, minimise fasting times, start IV fluids preoperatively if necessary

Anaesthesia for Repaired Lesion
- variety of surgical options depending on severity and details of lesion
- mild cases with good or acceptable right ventricular function: biventricular repair possible: tricuspid valve reconstruction/ repair/ replacement
- arrhythmia may persist → check ECG
- check Echo
  - persistent right ventricular dysfunction?
  - valve repair: residual stenosis or regurgitation?
- severe cases: single ventricle pathway with a 3-stage repair: Blalock-Taussig shunt (Chap. 36) → Glenn shunt (Chap. 38) → Total cavo-pulmonary connection TCPC (Chap. 44)

Additional Reading

1. Ross FJ, Latham GJ, Richards M, Geiduschek J, Thompson D, Joffe D. Perioperative and anesthetic considerations in Ebstein's anomaly. Semin Cardiothorac Vasc Anesth. 2016 Mar;20(1):82–92.
2. Dearani JA, Mora BN, Nelson TJ, Haile DT, O'Leary PW. Ebstein anomaly review: what's now, what's next? Expert Rev. Cardiovasc Ther. 2015 Oct 3;13(10):1101–9.
3. Hetzer R, Hacke P, Javier M, Miera O, Schmitt K, Weng Y, et al. The long-term impact of various techniques for tricuspid repair in Ebstein's anomaly. J Thorac Cardiovasc Surg. 2015 Nov;150(5):1212–9.
4. Kumar TKS, Boston US, Knott-Craig CJ. Neonatal Ebstein anomaly. Semin Thorac Cardiovasc Surg. 2017;29(3):331–7.
5. Morray B. Preoperative physiology, imaging, and management of Ebstein's anomaly of the tricuspid valve. Semin Cardiothorac Vasc Anesth. 2016 Mar;20(1):74–81.

Chapter 20
Heterotaxy Syndrome: Left Atrial Isomerism, Polysplenia Syndrome

Abstract Left atrial isomerism, also called Polysplenia syndrome, is one of the heterotaxy syndromes. It is a failure to differentiate into the right/left sidedness of the body, involving all internal organs. The cardiac anatomy can be highly variable, with conduction tissue abnormalities, septal defects, either atrial or ventricular, problems with valves, abnormal venous return, either systemic (vena cavae) or pulmonary, abnormal ventricular outflow and arteries, such as transposition of the great arteries or double outlet ventricles.

Anaesthetic management depends on the details of the lesion.

Keywords Heterotaxy syndrome · Left atrial isomerism · Polysplenia syndrome · PAPVD · Complex cardiac lesion · Arrhythmia · Paediatric cardiac anaesthesia · Congenital cardiac lesion.

Heterotaxy syndromes are a failure of differentiation into the left/right sidedness of the body, resulting in the malformationof several internal organs. It is a rare syndrome, but the complex anatomy will result in multiple operations for both cardiac and non-cardiac issues. Anaesthetic management depends on the individual lesion.

- Left atrial isomerism = two left sides of the body
- Non-cardiac:
  - Bilateral 2-lobed lung
  - Gastrointestinal malformation 80%
  - Occasional absent gallbladder
  - Symmetrical liver (usually midline)
  - Multiple spleens

Fig. 20.1 Left atrial isomerism with atrial septal defect and partial anomalous pulmonary venous return: right pulmonary artery to superior vena cava

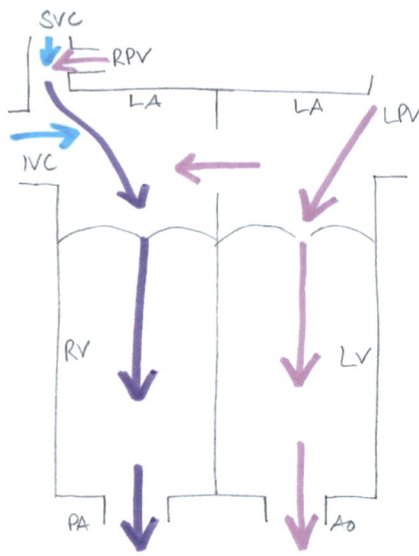

- Cardiac: very variable anatomy
    - Absent hepatic segment of IVC
    - Normal pulmonary venous return 50%
    - Partial anomalous pulmonary venous drainage PAPVD: 50% (Fig. 20.1)
        - right pulmonary veins to right sided atrium
        - left pulmonary veins to left sided atrium
    - Bilateral left atria, no sinus node
    - Single atrium or atrial septal defect ASD
    - Normal atrioventricular valves 50%, single atrioventricular valve 15%
    - two ventricles 80%, double outlet right ventricle DORV 20%
    - Ventricular septal defect VSD 65% (Fig. 20.2)
    - Transposition of the great arteries TGA 15%
    - Normal pulmonary valve 60%, pulmonary stenosis or atresia 40%
- Clinical:
    - Symptoms depend on individual lesion: congestive heart failure, cyanosis, arrhythmia, etc.
    - ECG: superior QRS axis, arrhythmia, ventricular hypertrophy
    - Chest x-ray: normal
    - Echo: details of lesion: septal defects with shunt direction and volume, anatomy of pulmonary veins, valve anatomy (stenosis, regurgitation?), function/contractility

Fig. 20.2 Left atrial isomerism with atrial and ventricular septal defect and partial anomalous pulmonary venous return: right pulmonary artery to superior vena cava

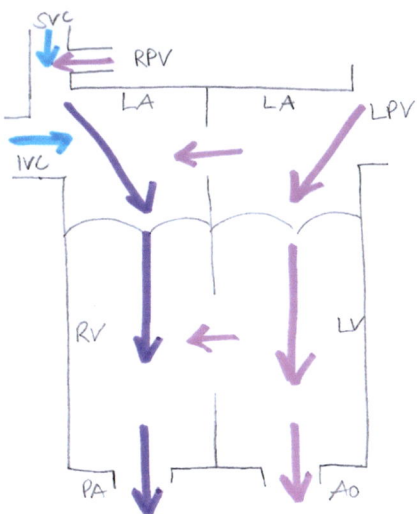

Anaesthetic Management
- Dependent on individual anatomy
- Maintain preload, keep well hydrated
- Ventricular septal defect VSD and pulmonary stenosis (Tetralogy type):
  - decrease pulmonary vascular resistance: FiO2 > 21%, mild hyperventilation
  - avoid decrease in systemic vascular resistance: fluids, vasopressors
- Pulmonary atresia: duct-dependent lesion:
  - Prostaglandin infusion
  - L → R Shunt
    - Avoid decrease in pulmonary vascular resistance: FiO2 21%, mild hypoventilation
    - Avoid increase in systemic vascular resistance: good analgesia and anaesthesia
- Arrhythmia: have pacing equipment available
- Cyanotic patients develop secondary erythrocytosis (Chap. 4):
  - coagulation disorder, making them prone to both thrombosis and bleeding
  - risk of acute hyperviscosity syndrome when dehydrated
  - keep well hydrated, minimise fasting times, start IV fluids preoperatively if necessary

Anaesthesia for Repaired Lesion
- Dependent on anatomy:
  - Full repair of cardiac lesions
  - Partial anomalous venous drainage: attachment of anomalous blood vessels to left sided atrium
  - Pulmonary artery band (Chap. 41), then ventricular septal defect repair
  - Double Outlet Right Ventricle DORV (Chap. 18): if full repair not possible: single ventricle pathway with a 3-stage repair:
    - Blalock-Taussig shunt (Chap. 36)
    - Glenn shunt (Chap. 38)
    - Total cavo-pulmonary connection TCPC (Chap. 44)
- Pacemaker for arrhythmia
- Check Echo for stenosis/ obstruction, residual shunts, ventricular function, etc.

Additional Reading

1. Kerai S, Gaba P, Gupta L, Saxena K. Anaesthetic management of a child with unrepaired complete atrioventricular canal defect, double outlet ventricle and pulmonary stenosis for non-cardiac surgery. Indian J Anaesth. 2022;66(18):342.
2. Simha P, Patel M, Jagadeesh A. Anesthetic implications of total anomalous systemic venous connection to left atrium with left isomerism. Ann Card Anaesth. 2012;15(2):134.
3. Ortega-Zhindón DB, Calderón-Colmenero J, García-Montes JA, Sandoval JP, Minakata-Quiroga MA, Cervantes-Salazar JL. Cardiac surgery in patients with atrial isomerism: long-term results and outcomes. J Card Surg. 2021 Dec;36(12):4476–84.
4. Greaney D, Honjo O, O'Leary JD. The single ventricle pathway in paediatrics for anaesthetists. BJA Educ. 2019 May;19(5):144–50.
5. Ortega-Zhindón DB, Flores-Sarria IP, Minakata-Quiróga MA, Angulo-Cruzado ST, Romero-Montalvo LA, Cervantes-Salazar JL. Atrial isomerism: A multidisciplinary perspective. Arch Cardiol Mex. 2021 Dec;91(4):470–9.

Chapter 21
Heterotaxy Syndrome: Right Atrial Isomerism, Asplenia Syndrome

Abstract Right atrial isomerism, also called Asplenia syndrome, is one of the heterotaxy syndromes. It is a failure to differentiate into the right/left sidedness of the body, involving all internal organs. The cardiac anatomy can be very variable, with abnormal venous return, both systemically and pulmonary (total anomalous pulmonary venous drainage TAPVD), conduction problems with two sinus nodes with atrial tachyarrhythmias, frequently complete atrio-ventricular septal defect with a common atrio-ventricular valve, pulmonary stenosis or atresia, etc.

Anaesthetic management depends on the details of the lesion.

Keywords Heterotaxy syndrome, right atrial isomerism · Single ventricle · Asplenia syndrome, TAPVD · Shunt lesion · Mixing lesion, cyanosis · Complex cyanotic lesion · Paediatric cardiac anaesthesia · Congenital cardiac lesion

Heterotaxy syndromes are a failure of differentiation into the left/right sidedness of the body, resulting in the malformation of several internal organs. It is a rare syndrome, but the complex anatomy will result in multiple operations for both cardiac and non-cardiac issues. Anaesthetic management depends on the individual lesion.

- Right atrial isomerism = two right sides of the body
- Non-cardiac:
    - Bilateral three-lobed lungs
    - Gastrointestinal malformation/malrotation
    - Midline symmetrical liver
    - No spleen

- Cardiac:
  - Bilateral superior vena cava, normal inferior vena cava (left-sided in 35%)
  - Total anomalous pulmonary venous drainage TAPVD: 80–90% (extracardiac 75%), often with pulmonary valve obstruction
  - Bilateral right atria, bilateral sinus node
  - Atrial septal defect ASD
  - Single atrioventricular valve 90% (Fig. 21.2)
  - Single ventricle 50% (Fig. 21.2), two ventricles 50% (Fig. 21.1)
  - Transposition of the great arteries TGA 70%
  - Pulmonary stenosis 40%, pulmonary atresia 40%

- Clinical:
  - Cyanosis
  - ECG: superior QRS axis, abnormal or missing p-waves, ventricular hypertrophy
  - Chest x-ray: normal or slightly enlarged heart
  - Echo: details of lesions: pulmonary veins, septal defects with shunt volume and direction, valve anatomy (stenosis, regurgitation?), function/contractility

- **High risk anaesthesia**

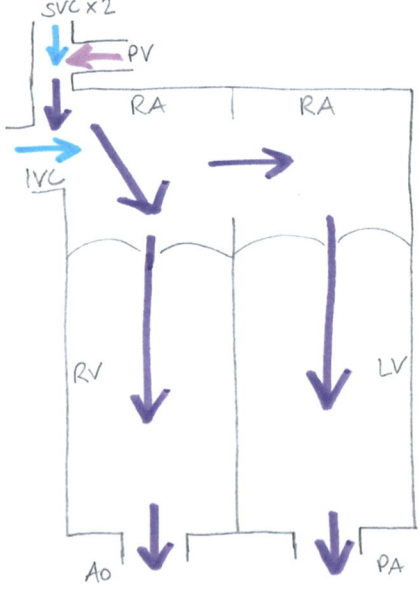

Fig. 21.1 Right atrial isomerism with bilateral SVC, total anomalous pulmonary venous drainage TAPVD, atrial septal defect ASD, two ventricles, transposition of the great arteries TGA

Fig. 21.2 Right atrial isomerism with bilateral SVC, total anomalous pulmonary venous drainage TAPVD, common atrium/ large ASD, one atrio-ventricular valve, one ventricle, transposition of the great arteries TGA

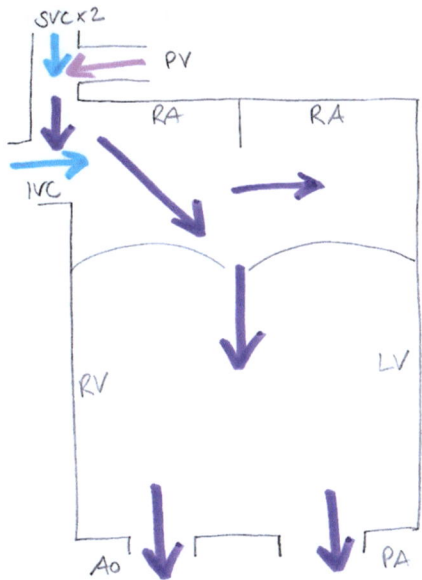

Anaesthetic Management
- Dependent on details of lesions
- Chronic cyanosis causes secondary erythrocytosis (Chap. 4):
  - coagulation disorder, prone to both thrombosis and bleeding
  - risk of acute hyperviscosity syndrome when dehydrated
- keep well hydrated, minimise fasting times, start IV fluids preoperatively if necessary
- Mixing single ventricle: balance pulmonary and systemic vascular resistance for optimal saturations and perfusion pressure
- In pulmonary stenosis: decrease pulmonary vascular resistance to promote pulmonary blood flow: FiO2 > 21%, mild hyperventilation
- Pulmonary atresia: duct-dependent:
  - Prostaglandin infusion
  - L → R shunt:
    - avoid decrease in pulmonary vascular resistance with FiO2 21% and mild hypoventilation
    - avoid increase in systemic vascular resistance with good anaesthesia and analgesia
- Might need inotropes for impaired function
- Arrhythmia possible → have pacing equipment available

Anaesthesia for the Repaired Lesion
- 2 ventricles: full repair might be possible
- Persistent arrhythmia? Pacemaker?
- If single ventricle: 3-stage repair:
    - Blalock-Taussig shunt (Chap. 36)
    - Glenn shunt (Chap. 38)
    - Total cavo-pulmonary connection TCPC (Chap. 44)

Additional Reading

1. Kerai S, Gaba P, Gupta L, Saxena K. Anaesthetic management of a child with unrepaired complete atrioventricular canal defect, double outlet ventricle and pulmonary stenosis for non-cardiac surgery. Indian J Anaesth. 2022;66(18):342.
2. Ortega-Zhindón DB, Calderón-Colmenero J, García-Montes JA, Sandoval JP, Minakata-Quiroga MA, Cervantes-Salazar JL. Cardiac surgery in patients with atrial isomerism: long-term results and outcomes. J Card Surg. 2021 Dec;36(12):4476–84.
3. Greaney D, Honjo O, O'Leary JD. The single ventricle pathway in paediatrics for anaesthetists. BJA Educ. 2019 May;19(5):144–50.
4. Ortega-Zhindón DB, Flores-Sarria IP, Minakata-Quiróga MA, Angulo-Cruzado ST, Romero-Montalvo LA, Cervantes-Salazar JL. Atrial isomerism: a multidisciplinary perspective. Arch Cardiol Mex. 2021 Dec;91(4):470–9.
5. Bansal S, Aron RA. Anesthetic considerations for non-cardiac surgery in an adult patient with right atrial isomerism. Grad Med Educ Res J. 2021;3(1):33.

Chapter 22
Hypoplastic Left Heart Syndrome

Abstract Hypoplastic left heart syndrome is an underdevelopment of the left ventricle and the left ventricular outflow tract, including valves. This mixing lesion is duct-dependent for systemic blood flow and can rapidly deteriorate to low cardiac output syndrome and congestive heart failure with metabolic acidosis. Decreasing pulmonary vascular resistance PVR results in pulmonary overcirculation and decreased systemic blood flow.

Anaesthetic management of the R → L shunt at ductal level means avoiding a decrease in PVR and avoiding an increase in systemic blood pressure.

Keywords HLHS · Hypoplastic left heart syndrome · Single ventricle complex cyanotic lesion · Duct-dependent, arrhythmias · Mixing lesion shunt lesion · Paediatric cardiac anaesthesia · Congenital cardiac lesion

Hypoplastic left heart syndrome HLHS, is rare, but severe, about 1% of all congenital heart disease. This is a cyanotic heart lesion and duct-dependent for systemic blood flow and oxygenation. Anaesthetic management mirrors the complexity of the lesion.

- Spectrum/ variety of underdevelopment of left ventricle and left outflow tract, including valves (see Fig. 22.1)

    - small left ventricle with hypoplastic aortic arch
    - Mitral valve abnormality, small annulus, stenosis or atresia
    - Aortic valve abnormality, small annulus, stenosis or atresia
    - 10% ventricular septal defect VSD
    - May have Coarctation as well
    - May have small coronary arteries
    - about 30% associated with brain abnormalities like corpus callosum agenesis or microcephaly

Fig. 22.1 Hypoplastic left heart, duct-dependent

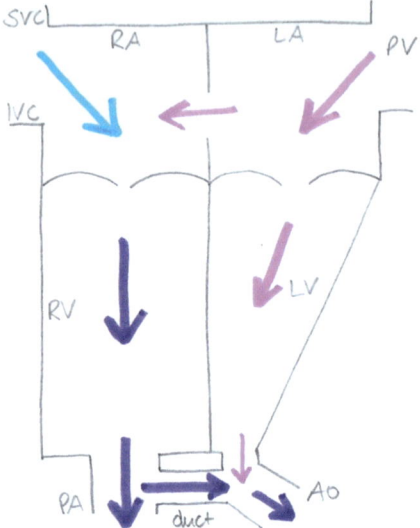

- If inadequate mixing (no atrial or ventricular septal defect or if defect is restrictive), an emergency balloon atrial septostomy (Rashkind procedure, Chap. 41) needs to be done
- Clinically:
  - Cyanosis (mixing lesion)
  - Low cardiac output syndrome, congestive heart failure, metabolic acidosis, increasing pulmonary blood flow due to decreasing pulmonary vascular resistance in the first few weeks of life
  - ECG: right ventricular hypertrophy, right heart volume loaded
  - Chest x-ray: pulmonary congestion/ oedema
  - Echo: details of lesion: septal defects and shunt volume (restrictive?), coronary anatomy and blood flow, right atrial dilatation?, additional abnormalities
- **High risk anaesthesia**

Anaesthetic Management
- Chronic cyanosis causes secondary erthrocytosis (Chap. 4):
  - coagulation disorder, making them prone to both thrombosis and bleeding
  - risk of acute hyperviscosity syndrome when dehydrated
  - keep well hydrated, minimise fasting times, start IV fluids preoperatively if necessary

- Duct-dependent for systemic blood flow:
    - Prostaglandin infusion
    - R → L shunt:
        - avoid decreasing pulmonary vascular resistance: FiO2 21%, mild hypoventilation
        - avoid increasing systemic vascular resistance: good anaesthesia and analgesia
- Mixing lesion
    - Avoid decreasing pulmonary vascular resistance: FiO2 21%, mild hypoventilation
    - Avoid increasing systemic vascular resistance: good anaesthetic, analgesia, vasodilatation if needed
- Right atrial dilatation due to volume load might cause arrhythmias
- Might need inotropes for improving ventricular function

Anaesthesia for Repaired Lesion
- Norwood operation (first week of life): arch repair, proximal pulmonary artery connected to aorta, right ventricular to pulmonary artery shunt (Sano) or Blalock-Taussig shunt (Chap. 40)
- If Norwood is not possible: Hybrid procedure (Chap. 39)
- Glenn shunt (3–6 months of life): takedown of BT shunt or Sano and connection of superior vena cava to pulmonary artery (Chap. 38)
- Total cavo-pulmonary connection TCPC/ Fontan circulation (3–7 years): connection of inferior vena cava to pulmonary artery (Chap. 44)

Additional Reading

1. Christensen RE, Gholami AS, Reynolds PI, Malviya S. Anaesthetic management and outcomes after noncardiac surgery in patients with hypoplastic left heart syndrome: a retrospective review. Eur J Anaesthesiol. 2012 Sep;29(9):425–30.
2. Lok T, Winch P, Naguib A. Perioperative management of a child with hypoplastic left heart syndrome following the hybrid stage I procedure presenting for laparoscopic gastrostomy tube placement. Pediatr Anesth Crit Care J. 2017 Jan 24;5(1):24–30.
3. Greaney D, Honjo O, O'Leary JD. The single ventricle pathway in paediatrics for anaesthetists. BJA Educ. 2019 May;19(5):144–50.

Chapter 23
Mitral Valve

Abstract This chapter describes mitral valve lesions: mitral valve stenosis is rare as an isolated defect, with clinical symptoms dependent on the degree of stenosis. Mild stenosis is asymptomatic, moderate to severe stenosis presents with dyspnoea due to pulmonary hypertension and right heart failure. Tachycardia should be avoided and inotropes might be needed for impaired function.

Mitral valve regurgitation or prolapse is rare as a isolated defect. Symptoms depend on severity of regurgitation and associated lesions. Anaesthetic management aims to promote forward flow by carefully reducing systemic vascular resistance while maintaining coronary perfusion pressures and to avoid bradycardia.

Keywords Mitral valve · Obstructive lesion · Regurgitant lesion · Valve lesion · Valvular lesion · Paediatric cardiac anaesthesia · Congenital cardiac lesion

23.1 Mitral Valve Stenosis

Mitral stenosis is rarely an isolated defect. It is usually part of a more complex lesion with additional issues. As an isolated defect it is most common as a complication after rheumatic fever, with a 5–10 year delay.

- Long term pathology:
  - Valve obstruction causes an increase in left atrial pressure
    - pulmonary congestion, then pulmonary hypertension PHT
    - increase in right ventricular pressure
    - tricuspid regurgitation
    - increase in right atrial pressure → Arrhythmia
  - Progressively impaired function, heart failure
- Clinical:
  - Dyspnoea, first on exertion, then at rest
  - Palpitations
  - Murmur
  - ECG: left atrial hypertrophy, right ventricular hypertrophy
  - Chest x-ray: pulmonary congestion/ oedema
  - Echo: defect, tricuspid regurgitation (late)

Anaesthetic Management
- Obstructive lesion: avoid tachycardia
- Pulmonary hypertension might benefit from decrease in pulmonary vascular resistance: FiO2 > 21%, mild hyperventilation
- Might need inotropes for impaired function
- Arrhythmia management

Anaesthesia for Repaired Lesion
- Check Echo for repair:
  - Residual stenosis? regurgitation?
  - Contractility/ function
  - If replacement valve: calcification? Structural valve deterioration?
  - Additional lesions?
- Check local/national guidelines regarding Endocarditis prophylaxis
- Pulmonary hypertension might persist
- Re-stenosis possible
- Valve replacement with biological or artificial valve: regular anticoagulation? Bridging with heparin needed?

23.2 Systolic Anterior Motion of the Mitral Valve

The systolic anterior motion of the mitral valve SAM is a special form of left ventricular outflow tract obstruction at mitral level.

- Usually associated with hypertrophic cardiomyopathy and abnormalities of the mitral valve or its subvalvar apparatus, but can happen in healthy hearts with severe hypovolaemia

- Mitral valve leaflet gets pulled into left ventricular outflow tract, resulting in obstruction
- Dynamic obstruction, depending on contractility and volume status
- Depending on anatomical factors can cause regurgitation as well
- Clinical: presents like left ventricular outflow tract obstruction
  - Dyspnoea
  - Low cardiac output syndrome LCOS
  - Might have murmur
  - ECG: ventricular hypertrophy
  - Echo: systolic anterior motion of mitral valve leaflet, mitral valve abnormality, sub-mitral apparatus abnormality with short cordae or papillary muscles, hypertrophic cardiomyopathy

Anaesthetic Management
- avoid tachycardia, betablockers are beneficial
- Hypovolaemia is detrimental: keep fasting times to a minimum, preoperative IV fluids if necessary, keep well filled
- decrease contractility: anaesthetic drugs with negative inotropy are beneficial

23.3 Mitral Valve Regurgitation

Mitral valve regurgitation is rare as an isolated defect, it is usually part of a more complex lesion.

- Can be due to dysplastic valve/ sub-valvar apparatus with chordae and papillary muscles
- can be secondary to left ventricular dilatation
- Clinical:
  - Dyspnoea, fatigue, low cardiac output failure LCOS
  - Late problems: congestive heart failure, pulmonary hypertension
  - Murmur
  - ECG: severe cases: hypertrophy of left atrium and ventricle, atrial fibrillation
  - Chest x-ray: enlarged left atrium and ventricle, pulmonary congestion/ plethora
  - Echo: valve defect, function/ contractility

Anaesthetic Management
- Regurgitant lesion: avoid bradycardia
- Maintain preload
- severe cases: might need inotropes
- Pulmonary hypertension might benefit from decrease in pulmonary vascular resistance: FiO2 > 21%, mild hyperventilation
- Arrhythmia management

Anaesthesia for Repaired Lesion
- Might have repair or valve replacement
- Check Echo for repair:
    - Residual problems?
    - Contractility/ function
    - If valve replacement: calcification? Structural valve deterioration?
    - Additional lesions?
- Check local/national guidelines regarding Endocarditis prophylaxis
- might have persistent arrhythmia → prone to arrhythmia due to ventriculotomy
- Valve replacement with biological or artificial valve: regular anticoagulation? Bridging with heparin needed?

23.4 Mitral Valve Prolapse

A mitral valve prolapse is a common lesion, affecting about 2–3% in the general population. It can be sporadic or familial and is more common in females. Valve prolapse may progress to regurgitation, with symptoms depending on the severity and additional complications such as arrhythmias.

- Congenital abnormality of mitral valve leaflets and chordae tendinae
- Can affect one or both mitral leaflets
- Dysplasia of the leaflet causes it to billow or prolapse into the left atrium in systole
- Associated with connective tissue disorders, e.g. Marfan's syndrome, Ehler-Danlos, etc. and chest wall deformities like pectus excavatum or scoliosis
- Clinical:
    - Mostly asymptomatic
    - Chest pain, palpitations, arrhythmia
    - Progressive disease: congestive heart failure, sudden cardiac death SCD (due to ventricular arrhythmias)
    - Murmur: stronger with manoeuvres that increase left ventricular volume, like Valsalva manoeuvre, squatting, etc.
    - ECG: normal. Might have supraventricular tachycardia, atrioventricular blocks, right bundle branch block
    - Chest x-ray: normal, might have left atrial enlargement in late stage
    - Echo: details of lesion: one or both leaflets involved, severity of regurgitation, contractility, additional lesions?

Anaesthetic Management
- Be aware of arrhythmias: pacing/ defibrilator available and ready
- Mitral regurgitation: avoid bradycardia
- Maintain good preload

- Might need inotropes for function
- In case of connective tissue disorders: careful attention to positioning

Anaesthesia for Repaired Lesion
- Arrhythmia can persist
- Check Echo for repair:
    - Residual issues? Good repair?
    - Contractility/ function
    - If valve replacement: calcification? Structural valve deterioration?
    - Additional lesions?
- Valve replacement: anticoagulation medication?
- Check local/national guidelines regarding Endocarditis prophylaxis

Additional Reading

1. Kabukcu HK, Sahin N, Kanevetci BN, Titiz TA, Bayezid O. Anaesthetic management of patient with Poland syndrome and rheumatic mitral valve stenosis: a case report. Ann Card Anaesth. 2005 Jul 1;8(2):145.
2. Paul A, Das S. Valvular heart disease and anaesthesia. Indian J Anaesth. 2017;61(9):721.
3. Grewal KS, Malkowski MJ, Piracha AR, Astbury JC, Kramer CM, Dianzumba S, et al. Effect of general anesthesia on the severity of mitral regurgitation by transesophageal echocardiography. Am J Cardiol. 2000 Jan;85(2):199–203.

Chapter 24
Pulmonary Atresia with Intact Ventricular Septum

Abstract Pulmonary atresia with intact ventricular septum is a duct-dependent complex cyanotic lesion. The complete obstruction at pulmonary valve level necessitates a R→L shunt at atrial level and a duct with a L→R shunt for pulmonary blood flow. High pressures in the right ventricle can cause coronary sinusoids, abnormal vascular connections between the right ventricle and the coronary system and a possible right ventricular pressure-dependent coronary circulation. Anaesthetic management is focussed on the ductal L→R shunt and the possible myocardial ischaemia due to the sinusoids.

Keywords Pulmonary atresia, cyanotic lesion · Shunt lesion, complex cyanotic lesion, coronary abnormalities · Duct-dependent, shunt lesion · Valve lesion, valvular lesion · Mixing lesion, coronary abnormalities · Paediatric cardiac anaesthesia · Congenital cardiac lesion

Pulmonary atresia is a rare lesion, less than 1% of congenital heart disease. It is duct-dependent for pulmonary blood flow and needs a R→L shunt at atrial level for decompressing the right heart by shunting the systemic venous return to the left atrium.

- RV size variable:
  - Tripartite: normal size inlet, infundibular, trabecular
  - Bipartite: inlet and infundibular
  - Monopartite: just inlet

- Needs atrial septal defect ASD or patent foramen ovale PFO and duct for survival (Fig. 24.1)
- High right ventricular pressure → coronary sinusoids

Fig. 24.1 Pulmonary atresia with intact ventricular septum, duct-dependent, atrial septal defect and right ventricular sinusoids

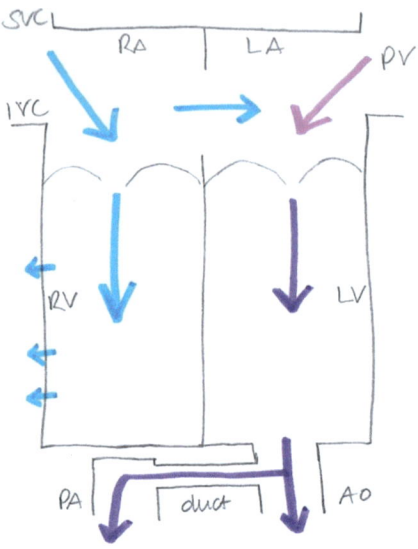

- Coronary perfusion might be right ventricular pressure dependent
- Clinical:
  - Cyanotic, unwell
  - Low cardiac output syndrome LCOS, congestive heart failure CHF
  - Murmur of atrial septal defect and duct
  - ECG: normal or left ventricular hypertrophy, left atrial hypertrophy
  - Chest x-ray: normal or cardiomegaly
  - Echo: lesion, atrial shunt needs to be unrestricted, coronary anatomy (sinusoids?), function/ contractility, ductal patency
- **High risk anaesthetic**

Anaesthetic Management
- Chronic cyanosis causes secondary erythrocytosis (Chap. 4):
  - coagulation disorder, making them prone to both thrombosis and bleeding
  - risk of acute hyperviscosity syndrome when dehydrated
- keep well hydrated, minimise fasting times, start IV fluids preoperatively if necessary
- If atrial shunt is restrictive: emergency balloon atrial septostomy (Rashkind procedure, Chap. 41)
- Duct-dependent
  - Prostaglandin infusion
  - L→R shunt at ductal level:
    - Avoid decrease in pulmonary vascular resistance: FiO2 21%, mild hypoventilation

- Avoid increase in systemic vascular resistance: good anaesthesia and analgesia
- Balance PVR and SVR for optimal blood pressure and oxygen delivery
- Coronary sinusoids with right ventricular pressure dependent coronary circulation: high mortality

Anaesthesia for Repaired Lesion
- Problems with coronary blood flow due to abnormal anatomy can persist
- Dependent on right ventricular size for a 2-ventricle or single ventricle repair
- Right ventricle tri- or bipartite: 2-ventricle repair: either pulmonary valvotomy/homograft with right ventricle to pulmonary artery conduit or Blalock-Taussig shunt (Chap. 36)
- Right ventricle monopartite, with or without sinusoids: single ventricle pathway: 3-stage repair:
  - Initial ductal stent (Chap. 41) or Blalock-Taussig shunt (Chap. 36)
  - Second stage: Glenn shunt (Chap. 38)
  - Third stage: Total cavo-pulmonary connection TCPC (Chap. 44)

Additional Reading

1. Wright LK, Knight JH, Thomas AS, Oster ME, St Louis JD, Kochilas LK. Long-term outcomes after intervention for pulmonary atresia with intact ventricular septum. Heart. 2019 Jul;105(13):1007–13.
2. Gleich S, Latham GJ, Joffe D, Ross FJ. Perioperative and anesthetic considerations in pulmonary atresia with intact ventricular septum. Semin Cardiothorac Vasc Anesth. 2018 Sep;22(3):256–64.
3. Chikkabyrappa SM, Loomba RS, Tretter JT. Pulmonary atresia with an intact ventricular septum: preoperative physiology, imaging, and management. Semin Cardiothorac Vasc Anesth. 2018 Sep;22(3):245–55.

Chapter 25
Partial Anomalous Pulmonary Venous Drainage

Abstract Partial anomalous pulmonary venous drainage (PAPVD) connects one or more pulmonary veins to the right sided circulation, resulting in pulmonary overcirculation due to a L→R shunt. Patients might be asymptomatic, with the diagnosis being an incidental finding, depending on how many veins are involved. Late stage problems are right atrial dilatation causing arrhythmias and pulmonary hypertension with right ventricular failure due to persistent volume overload.

Anaesthetic management depends on shunt volume and complications such as pulmonary hypertension or arrhythmias.

Keywords Partial anomalous pulmonary venous drainage, PAPVD · Vascular lesion, arrhythmia · Shunt lesion, pulmonary hypertension · Paediatric cardiac anaesthesia · Congenital cardiac lesion

Partial anomalous venous drainage is rare at less than 1% of congenital heart disease. The pulmonary venous return to the right side of the heart causes increased pulmonary blood flow with a pulmonary volume load. Symptoms and progression of disease depend on the number of anomalous veins.

- One or more pulmonary veins drain to right atrium or venous contributories (Fig. 25.1): superior or inferior vena cava, coronary sinus, innominate vein, etc.
- Right pulmonary veins involved twice as often as left pulmonary veins
- Right pulmonary veins → superior vena cava or inferior vena cava
- usually associated with atrial septal defect ASD (80–90% of cases)
- Left pulmonary veins → coronary sinus, innominate vein
- Anomalous veins might have broncho-pulmonary sequestration as well
- L→R shunt

- Clinical
  - Mostly asymptomatic
  - > 1 vein: large shunt
  - Murmur: might have atrial septal defect murmur (usually associated with right pulmonary veins)
  - ECG: normal, might have right ventricular hypertrophy or right bundle branch block RBBB
  - Chest x-ray: might show Scimitar syndrome (upper right pulmonary veins to inferior vena cava, Fig. 25.2)

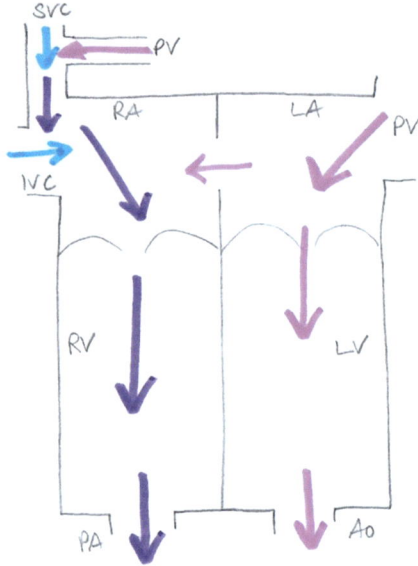

Fig. 25.1 Partial anomalous pulmonary venous drainage PAPVD: 1-3 pulmonary veins draining into the superior vena cava

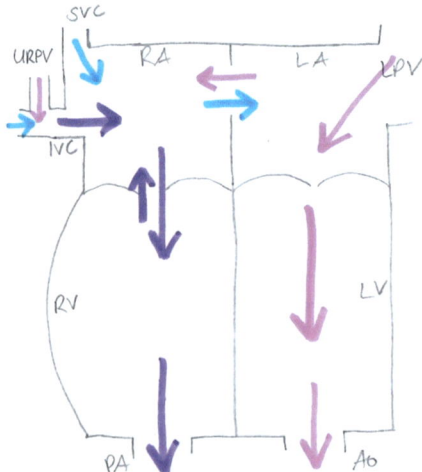

Fig. 25.2 Partial anomalous pulmonary venous drainage PAPVD: the upper right pulmonary vein URPV drains into the inferior vena cava IVC: Scimitar syndrome

- Echo: pulmonary veins not draining into left atrium, volume loaded right atrium and ventricle, might have patent foramen ovale due to increased pressure in right atrium
- Late problems: pulmonary hypertension, pulmonary vascular disease → cyanosis, exertional dyspnoea, syncope, right ventricular dilatation and dysfunction (see Fig. 25.2) with tricuspid regurgitation

Anaesthetic Management
- Usually stable, unless severe pulmonary hypertension PHT or right ventricular failure
- L→R shunt, but entirely within the pulmonary circulation
    - Keep pulmonary vascular resistance at pre-op levels: FiO2 21%, mild hypoventilation
    - Oxygen (FiO2 >21%) will lower pulmonary vascular resistance and might be beneficial in pulmonary hypertension PHT
- Careful management of impaired right ventricular function and/ or pulmonary hypertension
- Prone to atrial arrhythmias due to right atrial dilatation

Anaesthesia for the Repaired Lesion
- Usually good repair
- Might have stenosis at anastomosis site → obstruction causing pulmonary hypertension PHT
- Right ventricular impairment might persist

Additional Reading

1. Lewis RA, Billings CG, Bolger A, Bowater S, Charalampopoulos A, Clift P, et al. Partial anomalous pulmonary venous drainage in patients presenting with suspected pulmonary hypertension: a series of 90 patients from the ASPIRE registry. Respirology. 2020 Oct;25(10):1066–72.

Chapter 26
Patent Ductus Arteriosus

Abstract Patent or persistent ductus arteriosus is a vascular connection between the pulmonary artery and the aorta, causing a L→R shunt. The ductus is a normal part of the antenatal circulation and usually closes in the first few days of life. Persistence is more common in premature babies. Pulmonary vascular resistance decreases in the first few weeks of life, increasing shunt volume and therefore pulmonary overcirculation, resulting in tachypnoea and pulmonary oedema. Anaesthetic management aims to avoid a decrease in pulmonary vascular resistance with air and mild hypoventilation and avoid an increase in systemic blood pressure.

Keywords PDA · Patent ductus arteriosus · Vascular lesion · Shunt lesion · Pulmonary hypertension · Paediatric cardiac anaesthesia · Congenital cardiac lesion

A patent ductus arteriosus occurs in 5–10% of congenital heart disease and is more common in premature babies. It is a L→R shunt with overcirculation of the pulmonary system, leading to pulmonary hypertension and congestive heart failure.

- Ductus arteriosus is a normal part of fetal circulation, usually closes at birth or within first few days of life (Chap. 3)
- Connects the descending aorta to the main pulmonary artery (Fig. 26.1)
  - Antenatal: R→L shunt due to high pulmonary vascular resistance PVR
  - Postnatal: L→R shunt due to fall in pulmonary vascular resistance PVR → decreasing PVR over the first weeks of life increases shunt

Fig. 26.1 Patent or persistent ductus arteriosus

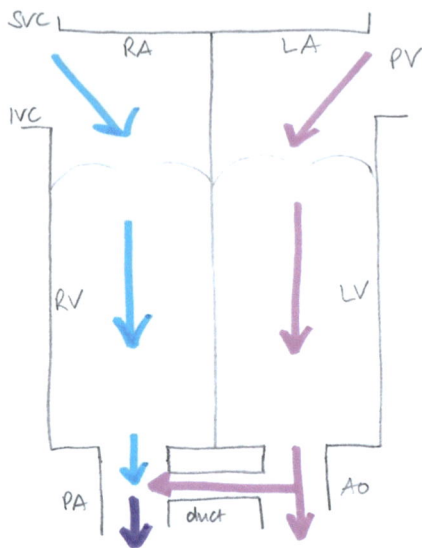

- Clinical:
  - Tachypnoea due to high pulmonary blood flow
  - Feeding problems, failure to thrive
  - Will develop into pulmonary hypertension and congestive heart failure
  - Murmur
  - ECG: normal or left ventricular hypertrophy, biventricular hypertrophy
  - Chest x-ray: normal or cardiomegaly, plethoric lungs
  - Echo: lesion, additional defects?

Anaesthetic Management
- L→R shunt
  - Avoid decrease in pulmonary vascular resistance: FiO2 21%, mild hypoventilation
  - Avoid increase in systemic vascular resistance: good anaesthesia and analgesia
- Careful fluid management
- Patients with congestive heart failure might need inotropes

Anaesthesia for Repaired Lesion
- Usually no long-term problems
- Additional lesions?

Additional Reading

1. Shinde S, Basantwani S, Tendolkar B. Anesthetic management of patent ductus arteriosus in adults. Ann Card Anaesth. 2016;19(4):750.
2. Janvier A, Martinez JL, Barrington K, Lavoie J. Anesthetic technique and postoperative outcome in preterm infants undergoing PDA closure. J Perinatol. 2010 Oct;30(10):677–82.

Chapter 27
Pulmonary Valve

Abstract This chapter describes pulmonary valve lesions: pulmonary stenosis can be an isolated lesion or part of a complex lesion. Mild cases are asymptomatic, moderate to severe stenosis causes dyspnoea, with arrhythmias and right ventricular failure in the late stages. Anaesthetic management aims to avoid tachycardia and manage arrhythmias. Inotropes might be needed in heart failure.

Pulmonary regurgitation is common after Tetralogy of Fallot's repair or balloon valvotomies. Mild cases are asymptomatic, moderate to severe cases present with dyspnoea and right heart failure. Anaesthetic management aims to promote forward flow by reducing pulmonary vascular resistance and to avoid bradycardia.

Keywords Pulmonary valve · Valve lesion · Valvular lesion · Regurgitant lesion · Duct-dependent · Obstructive lesion · Arrhythmias · Paediatric cardiac anaesthesia · Congenital cardiac lesion

27.1 Pulmonary Valve Stenosis

A pulmonary valve stenosis is relatively common, accounting for up to 10% of all congenital heart disease. It can occur in isolation or in combination with other lesions. Severe/ critical stenosis is high risk for right heart failure, arrhythmias and cyanosis due to a R → L shunt via the foramen ovale with high right atrial pressures.

- Types:
  - Valvular: 90%
  - Sub-valvular/ infundibular
  - Supra-valvular = main pulmonary artery stenosis or narrowing, small main pulmonary artery

- Right ventricular hypertrophy will develop depending on severity
- Severity grading:
    - mild: <35–40 mmHg (right ventricular pressure < 50% of left ventricular pressure)
    - moderate: 40–70 mmHg (RV pressure 50–75% of LV pressure)
    - severe: >70 mmHg (RV pressure > 75% of LV pressure)
- Right ventricle usually normal in size, but in congenitally critical pulmonary stenosis the right ventricle is hypoplastic
- Clinical:
    - Asymptomatic in mild to moderate cases
    - Exertional dyspnoea in moderate to severe cases with congestive heart failure and possible cyanosis in the end stages
    - Murmur
    - ECG: normal, right ventricular hypertrophy, tricuspid regurgitation and right atrial dilatation in severe cases, causing arrhythmia
    - Chest x-ray: normal, but prominent main pulmonary artery (post-stenotic dilatation)
    - Echo: lesion and degree of stenosis, ventricular function, tricuspid regurgitation?

Anaesthetic Management
- Stenosis/ obstruction of pulmonary blood flow
    - Keep pulmonary vascular resistance low: FiO2 > 21%, mild hyperventilation
    - Avoid decrease in systemic vascular resistance: fluids, vasopressors
    - Avoid tachycardia
- Maintain preload
- Maintain ductal patency in neonatal critical stenosis:
    - Prostaglandin infusion
    - L → R shunt via the duct: FiO2 21%, mild hypoventilation, avoid increase in systemic vascular resistance
- Severe stenosis: reduced cardiac output
    - right ventricular volume load → increase in right ventricular pressure and tricuspid regurgitation → right atrial volume load, then pressure increase: can cause R → L shunt through patent foramen ovale, causing cyanosis
    - increased right atrial pressure and dilatation can cause arrhythmias

Anaesthesia for Repaired Lesion
- Balloon valvotomy: usually first line treatment for severe or isolated stenosis, varying outcomes dependent on anatomy (dysplastic valve, small annulus) and age of patient: check Echo for:
  - Residual stenosis or re-stenosis
  - Pulmonary regurgitation
- Surgical repair/ replacement: usually good long-term outcomes
  - Surgical valvotomy/ transannular patch: good outcomes, but pulmonary regurgitation a common problem: check Echo for regurgitation, additional lesions, function/ contractility
  - Valve replacement: different options (homograft, bioprosthetic, mechanical)
    - Homograft/ bioprosthetic: no anticoagulation needed
    - Mechanical: usually on anticoagulation drugs
- Check Echo for repair:
  - Residual stenosis? regurgitation?
  - Contractility/ function
  - Replacement valve function: calcification? Structural valve deterioration?
  - Additional lesions?
- Check local/national guidelines regarding Endocarditis prophylaxis

27.2 Pulmonary Regurgitation

Pulmonary regurgitation is rare as an isolated lesion, usually after Tetralogy of Fallot's repair or balloon valvotomy. It causes right ventricular volume load which is usually well tolerated, but over time can cause right ventricular dilatation leading to dysfunction and failure, with tricuspid regurgitation and atrial arrhythmias.

- Clinical:
  - Mild to moderate: asymptomatic
  - Moderate to severe: exertional dyspnoea, syncope, ventricular tachycardia, arrythmia, congestive heart failure
  - Murmur
  - ECG: nothing specific
  - Chest x-ray: enlarged right ventricle, late sign cardiomegaly
  - Echo: lesion, regurgitant fraction, right ventricular volume index and function

Anaesthetic Management
- Regurgitant lesion:
  - Decrease pulmonary vascular resistance: $FiO_2 > 21\%$, mild hyperventilation
  - Avoid bradycardia
- Enhance right ventricular function with inotropes
- Maintain preload

Anaesthesia for Repaired Lesion
- Valve replacement: different options (homograft, bioprosthetic, mechanical): good long-term outcomes
  - Homograft/ bioprosthetic: no anticoagulation needed
  - Mechanical: usually on anticoagulation drugs
- Check Echo for repair:
  - Residual or additional lesion?
  - Contractility/ function
  - Replacement valve function: calcification? Structural valve deterioration?
- Check local/national guidelines regarding Endocarditis prophylaxis

Additional Reading

1. Sushma KS, Shaikh S. Anaesthetic management of pulmonary stenosis already treated with pulmonary balloon valvuloplasty. J Clin Diagn Res. 2014 Jan 12;8(1):193–4.

Chapter 28
Tetralogy of Fallot's

Abstract Tetralogy of Fallot's is a cyanotic heart disease with a ventricular septal defect, an overriding aorta, right ventricular outflow tract obstruction and a resulting right ventricular hypertrophy. The lesion is a combination of obstruction and a R→L shunt at ventricular level. Anaesthetic management of this lesion is to maintain preload, decrease pulmonary vascular resistance with oxygen and mild hyperventilation to improve forward flow through the right ventricular outflow tract obstruction and to avoid tachycardia. Inotropes might be needed for impaired function.

Keywords Tetralogy of Fallot's, TOF · Fallot's · Tetralogy · Ventricular septal defect · Shunt lesion · Obstructive lesion · Spells · Cyanosis · Paediatric cardiac anaesthesia · Congenital cardiac lesion

Tetralogy of Fallot's is the most common cyanotic heart disease, about 10% of all congenital heart disease. If the right ventricular outflow tract obstruction is mild, the pressures in the left ventricle are higher than in the right, resulting in a L→R shunt with no cyanosis, called a "pink Fallot's" (Fig. 28.2). Symptoms depend on the combination of the different lesion and their severity. Of particular concern are so-called "spells", hypercyanotic episodes caused by an increased R→L shunt at ventricular level. Anaesthetic management depends on the individual details of the lesion.

- "Tetralogy" (see Fig. 28.1):
  - Ventricular septal defect VSD
  - Aorta overriding the VSD
  - Right ventricular outflow tract obstruction RVOTO
  - Right ventricular hypertrophy RVH
    - spectrum of severity => variety of symptoms

© The Author(s), under exclusive license to Springer Nature Switzerland AG 2025
J. Scheffczik, *Paediatric Cardiac Anaesthesia*,
https://doi.org/10.1007/978-3-031-90330-4_28

Fig. 28.1 Tetralogy of Fallot's with R→L ventricular shunt

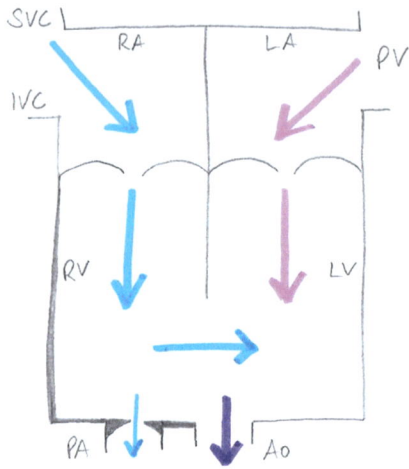

- VSD: usually perimembranous, shunt flow depending on RVOTO
  - Usually high right ventricular pressures: R→L shunt
  - In mild pulmonary stenosis: bidirectional shunt or L→R shunt
- RVOTO:
  - infundibular (muscular) 50% } both: 30%
  - Valvar: 10%
  - Pulmonary Atresia: 10%
- Aortic override:
  - <50%: Tetralogy of Fallot's
  - >50%: double outlet right ventricle
- Abnormal coronaries: 5%, right aortic arch 25%
- Clinical:
  - Cyanosis, tachypnoea
  - Hyper-cyanotic episodes: "spells"
  - Murmur from ventricular septal defect and pulmonary stenosis
  - ECG: right ventricular and atrial hypertrophy, might have biventricular hypertrophy in pink Fallot's
  - Chest x-ray: normal or boot-shaped heart
  - Echo: details of lesions, coronary anatomy

"Spells": hyper-cyanotic episodes due to increased R→L shunt

- Causes:
  - right ventricular outflow tract spasm
  - tachycardia (due to being upset or in pain)
  - increase in pulmonary vascular resistance (crying, screaming, defecation, etc)
  - decrease in systemic vascular resistance (dehydration, medication causing vasodilatation, etc)
- Treatment depends on cause, but multiple causes possible
- Tachycardia: anaesthesia/ analgesia (Morphine), calming an awake child, beta-blocker
- Increase in pulmonary pressures: FiO2 100%, mild hyperventilation in an anaesthetised child
- Decrease in systemic pressures: fluid bolus, vasopressors, folding legs up onto chest

Anaesthetic Management
- Chronic cyanosis causes secondary erythrocytosis (Chap. 4)
  - coagulation disorder, making them prone to both thrombosis and bleeding
  - risk of acute hyperviscosity syndrome when dehydrated
  - keep well hydrated, minimise fasting times, start IV fluids preoperatively if necessary
- premedication to prevent hyper-cyanotic spell caused by stress response to anaesthetic induction
- Combination of obstruction and shunt/mixing
- Obstruction and R→L shunt:
  - avoid increase in pulmonary vascular resistance: FiO2 >21%, mild hyperventilation
  - avoid decrease in systemic vascular resistance: fluids, vasopressors
  - maintain preload: fluids
  - avoid tachycardia and an increase in systemic vascular resistance: good anaesthesia and analgesia
- L→R shunt ("pink TOF", Fig. 28.2):
  - avoid decrease in pulmonary vascular resistance: FiO2 21%, mild hypoventilation
  - avoid increase in systemic vascular resistance: good analgesia and anaesthesia

Anaesthesia for Repaired Lesion
- Usually good repair with good long-term outcomes
- Coronary anatomy may determine type of repair
- Check Echo for:
  - Residual ventricular septal defect
  - Pulmonary regurgitation (common problem after pulmonary valvotomy / transannular patch repair)

Fig. 28.2 "Pink Tetralogy": minimal pulmonary stenosis, no R→L shunt, but a L→R shunt

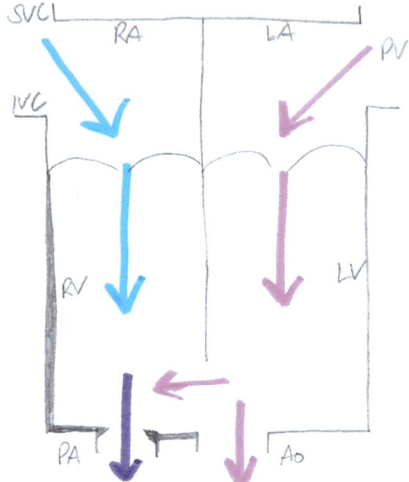

- Coronary anatomy and blood flow
- Contractility/ function
- Additional lesions

- Coronary abnormalities can cause persistent problems
- Small babies, in whom full repair is considered high risk, can have a right ventricular outflow tract RVOT stent inserted as a bridge to full repair

  - Check Echo for RVOT flow and gradient
  - RVOT stents will get epithelialised over time → re-stenosis → either balloon of the stent or full repair if possible
  - RVOT stents can cause arrhythmias

Additional Reading

1. Utomo MP, Vitraludyono R, Yupono K. Anaesthesia perioperative management in laparotomy procedure in neonates with tetralogy of fallot (ToF): a case study. EJMED. 2023 Feb 28;5(1):16–9.
2. Van Der Ven JPG, Van Den Bosch E, Bogers AJCC, Helbing WA. Current outcomes and treatment of tetralogy of Fallot. F1000Res. 2019 Aug 29;8:1530.
3. Dwivedi P, Kumar S, Ahmad S, Sharma S. Uncorrected tetralogy of Fallot's: anesthetic challenges. Anesth Essays Res. 2020;14(2):349.
4. Indumathi S, Kamble R, Adsule P, Bhalerao PM. Anesthetic management for transverse colostomy in a neonate with anorectal malformation and uncorrected tetralogy of fallot. J Res Innov Anesth. 2019 Dec 1;4(1):7–8.

Chapter 29
Tetralogy of Fallot's with Pulmonary Atresia (Pulmonary Atresia with Ventricular Septal Defect)

Abstract Tetralogy of Fallot's with pulmonary atresia is a duct-dependent cyanotic heart lesion. This is a combination of a ventricular septal defect, an overriding aorta, pulmonary atresia and right ventricular hypertrophy. There is a R→L shunt at ventricular level and a L→R shunt at ductal level to provide pulmonary blood flow. Symptoms depend on pulmonary arteries and additional collateral blood vessels from the aorta to the pulmonary circulation.

Anaesthetic management can be challenging and needs to avoid decreasing pulmonary vascular resistance and avoid increasing systemic blood pressure.

Keywords Pulmonary atresia, TOF · Tetralogy, Fallot's · MAPCAs · Dependent, ventricular septal defect · Obstructive lesion · Shunt lesion · Mixing lesion · Cyanosis · Complex cardiac lesion · Paediatric cardiac anaesthesia · Congenital cardiac lesion

Pulmonary atresia with a ventricular septal defect is a duct dependent lesion: 2–6% of Tetralogy of Fallot's patients have pulmonary atresia (Fig. 29.1). The lack of blood flow through the pulmonary artery can cause hypoplasia. Symptoms depend on the details of the individual lesion. Anaesthetic management mirrors the complexity of the lesion.

- Duct-dependent
- Pulmonary blood flow through duct and major aortopulmonary collateral arteries = MAPCAs
- Management and clinical picture dependent on pulmonary arteries size and confluence:
  - Confluent pulmonary arteries, normal size, 1–2 MAPCAs
  - Non-confluent, hypoplastic pulmonary arteries, multiple MAPCAs

Fig. 29.1 Tetralogy of Fallot's with pulmonary atresia, duct dependent. Also called pulmonary atresia with VSD. MAPCAs are major aorto-pulmonary collateral arteries: pathological blood vessels from aorta providing pulmonary blood supply, creating a L→R shunt

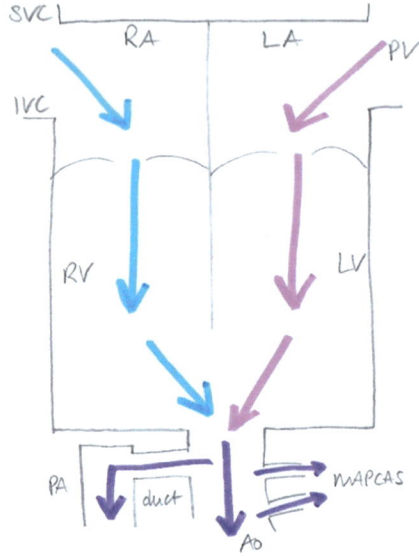

- Aneurysmal dilatation of pulmonary artery antenatally possible
  - compression of trachea and bronchi → will hinder development
- Clinical:
  - Cyanosis
  - Tracheomalacia: stridor, recurrent respiratory infections, etc
  - Murmur from ventricular septal defect and duct
  - ECG: right ventricular and atrial hypertrophy, arrhythmias
  - Chest x-ray: dilated main pulmonary artery and hilar pulmonary arteries, normal heart size
  - Echo: details of lesion, size and position of ventricular septal defect, aortic override, coronary anatomy, pulmonary artery anatomy
- **High risk anaesthesia**

Anaesthetic Management
- Mixing and L→R shunt at ductal level
- Treat MAPCAs as L→R shunt
  - Avoid decrease in pulmonary vascular resistance: FiO2 21%, mild hypoventilation
  - Avoid increase in systemic vascular resistance: good anaesthesia and analgesia
- Chronic cyanosis causes secondary erythrocytosis (Chap. 4)
  - coagulation disorder, making them prone to both thrombosis and bleeding
  - risk of acute hyperviscosity syndrome when dehydrated
- keep well hydrated, minimise fasting times, start IV fluids preoperatively if necessary

Anaesthesia for Repaired Lesion
- Surgical repair depends on details of anatomy
- Full repair can be done in one operation or staged, depending on anatomy
- Confluent, good size pulmonary arteries: full repair with a pulmonary valve homograft or bioprosthesis
- Unifocalisation of MAPCAs: all MAPCA origins are disconnected from aorta and anastomosed to the pulmonary artery as a single vessel
- Hypoplastic pulmonary arteries: right ventricle to pulmonary artery conduit ± unifocalisation ± ventricular septal defect closure
- Non-confluent, small pulmonary arteries with MAPCAs: RV to PA conduit, unifocalisation, VSD closure
- Check Echo:
  - Details of repair and pulmonary blood flow
  - Conduit: stenosis, narrowing, calcification
  - Residual ventricular septal defect?
  - If valve replacement: calcification, stenosis, regurgitation, structural deterioration?
  - Contractility/function
- Valve replacement: different options (homograft, bioprosthetic, mechanical)
  - Homograft/bioprosthetic: no anticoagulation needed
  - Mechanical: usually on anticoagulation drugs
- Pulmonary problems like tracheomalacia can persist even with full repair
- Check local/national guidelines regarding Endocarditis prophylaxis

Additional Reading

1. Herady MV, Ganigara A. Case of uncorrected and inoperable tetralogy of fallot with pulmonary atresia and major Aorto pulmonary collaterals with anorectal malformation-anesthetic management mimicking the pathophysiology in a limited cardiac resource centre. Ann Clin Case Rep. 2020;5:1831.

Chapter 30
Tricuspid Atresia (Hypoplastic Right Heart)

Abstract Tricuspid Atresia is a complex cyanotic lesion with a functionally single ventricle circulation. This is a mixing lesion with a R→L shunt at atrial level, a L→R shunt at ventricular and duct level. A transposition of the great arteries is present at 30% of cases and associated lesions are Coarctation and interrupted aortic arch.

Anaesthetic management depends on the details of the lesion. Most important are an unrestricted atrial shunt and to avoid overcirculating the lungs by avoiding a decrease in pulmonary vascular resistance and avoiding an increase in systemic blood pressure.

Keywords Tricuspid valve, tricuspid atresia · Valvular lesion, single ventricle · Valve lesion, mixing lesion · Obstructive lesion, hypoplastic right heart syndrome · Regurgitant lesion · Paediatric cardiac anaesthesia · Congenital cardiac lesion

Tricuspid atresia is a rare, but complex cyanotic lesion, about 1–3% of all congenital heart disease. It is reliant on an unrestricted R→L shunt on atrial level for survival and duct-dependent for pulmonary blood flow.

- Combination of tricuspid valve atresia, small right ventricle, ventricular septal defect, atrial septal defect, pulmonary stenosis or normal pulmonary valve, hypoplastic main pulmonary artery (see Fig. 30.1)
- 30% Transposition of the great arteries TGA
- Duct-dependent
- Associated with Coarctation or Interrupted Aortic Arch
- If restrictive atrial septum: emergency balloon septostomy (Rashkind procedure, Chap. 41)

Fig. 30.1 Tricuspid atresia with atrial septal defect, ventricular septal defect, small right ventricle, small pulmonary artery, duct-dependent

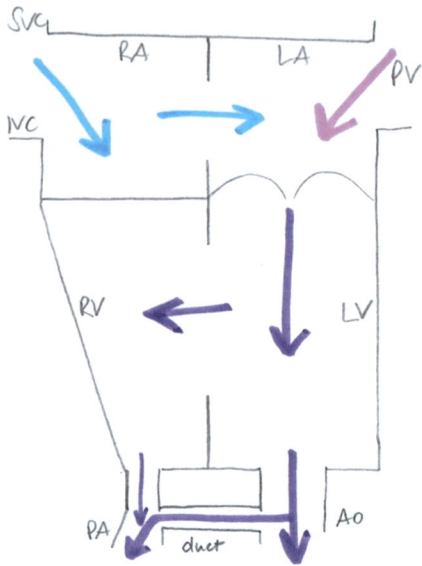

- Clinical:
  - Cyanosis
  - Congestive heart failure, pulmonary hypertension
  - Murmur from shunts and duct
  - ECG: "superior axis" right atrial hypertrophy, bi-atrial hypertrophy, left ventricular hypertrophy
  - Chest x-ray: Normal or slightly enlarged heart
  - Echo: details of lesion: atrial septal defect size (unrestricted?) ventricular septal defect size and shunt volume, pulmonary valve anatomy and gradient, transposition, additional lesions
- High risk anaesthesia

Anaesthetic Management
- Atrial septal defect needs to be unrestricted, as entire systemic venous return needs to shunt to left atrium
- L→R shunt at ductal level
  - Avoid decrease in pulmonary vascular resistance: FiO2 21%, mild hypoventilation
  - Avoid increase in systemic vascular resistance: good anaesthesia and analgesia
- Mixing at left atrial level: keep pulmonary and systemic blood pressures at preop levels and change as needed

- Chronic cyanosis causes secondary erythrocytosis (Chap. 4)
  - coagulation disorder, making them prone to both thrombosis and bleeding
  - risk of acute hyperviscosity syndrome when dehydrated
  - keep well hydrated, minimise fasting times, start IV fluids preoperatively if necessary

Anaesthesia for Repaired Lesion
- Full repair is usually not possible: single ventricle pathway
- Initial repair dependent on lesion:
  - TA and insufficient pulmonary blood flow: Blalock-Taussig shunt (Chap. 36)
  - TA/TGA/small VSD: Damus-Kaye-Stansel operation (Chap. 37) and BT shunt (Chap. 36)
  - TA/TGA/high pulmonary blood flow: Pulmonary artery band (Chap. 41)
- 2nd stage: Glenn shunt or Hemifontan (Chap. 38)
- 3rd stage: total cavo-pulmonary connection TCPC (Chap. 44)

Additional Reading

1. Yunus M, Saikia MK, Das SK, Karim HM, Mitra JK. Perioperative management of patients with tricuspid atresia and univentriclular congenital cardiac defect. Anaesth Pain Intensive Care. 2019;18:189–91.
2. Greaney D, Honjo O, O'Leary JD. The single ventricle pathway in paediatrics for anaesthetists. BJA Education. 2019;19(5):144–50.

Chapter 31
Tricuspid Valve

Abstract This chapter describes tricuspid valve lesions: tricuspid valve stenosis is rare as an isolated defect, with clinical symptoms dependent on the degree of stenosis. Anaesthetic management aims to avoid tachycardia and promote forward flow with decreasing pulmonary vascular resistance. Inotropes might be needed for impaired function and arrhythmias need to be treated.

Tricuspid valve regurgitation is rare as a isolated defect. Symptoms depend on severity of regurgitation and associated lesions. Anaesthetic management aims to promote forward flow by decreasing pulmonary vascular resistance and to avoid bradycardia.

Keywords Tricuspid valve · Valve lesion, valvular lesion · Obstructive lesion · Regurgitant lesion · Paediatric cardiac anaesthesia · Congenital cardiac lesion

31.1 Tricuspid Stenosis

Tricuspid valve stenosis is rare as an isolated lesion. Symptoms depend on severity of gradient and the increase of right atrial pressures, leading to arrythmias and a potential R→L shunt through the foramen ovale.

- Causes:
  - Congenitally dysplastic
  - Acquired (rheumatic fever, systemic lupus erythematosus, etc)
  - Iatrogenic after cardiac surgery or pacemaker insertion

- Pathophysiology: increased right atrial pressure:
  - Might reopen a patent foramen ovale, resulting in a R→L shunt
  - Long-standing increased right atrial pressures leads to right atrial dilatation, which causes arrhythmias
  - Reduced pulmonary blood flow
  - Increased central venous pressure CVP
  - Systemic congestion: peripheral oedema, ascites, etc
- Clinical:
  - Fatigue, signs of right heart failure, peripheral oedema
  - Mild cyanosis in case of R→L shunt
  - Murmur
  - ECG: right atrial enlargement, arrhythmia (atrial fibrillation)
  - Chest x-ray: right atrial enlargement
  - Echo: lesion, gradient of stenosis, right atrial dilatation?, reopened foramen ovale with R→L shunt?, additional lesions?

Anaesthetic Management
- Avoid tachycardia
- Promote forward flow by decreasing pulmonary vascular resistance with FiO2 >21% and mild hyperventilation
- Treat arrhythmias, either medication or cardioversion
- Chronic cyanosis causes secondary erythrocytosis (Chap. 4)
  - coagulation disorder, making them prone to both thrombosis and bleeding
  - risk of acute hyperviscosity syndrome when dehydrated
- keep well hydrated, minimise fasting times, start IV fluids preoperatively if necessary
- Might need inotropes

Anaesthesia for Repaired Lesion
- Outcome depends on underlying condition
- Either repair or valve replacement (homograft, bioprosthesis, mechanical/artificial valve)
- Arrhythmia may persist
- Valve replacement: anticoagulation management might need stopping or bridging with heparin might be needed
- Check local/national guidelines about endocarditis prophylaxis

31.2 Tricuspid Regurgitation

Tricuspid regurgitation is rare as an isolated lesion, it is most common as part of Ebstein's anomaly (Chap. 19). Regurgitation causes a volume-loaded right atrium, which over time causes arrhythmias and right heart failure.

- Causes:
  - Congenital, most commonly with Ebstein's anomaly
  - Acquired (infective endocarditis, rheumatic fever, etc)
  - Functional regurgitation: pathology of other lesions leads to dilatation of right heart and tricuspid valve annulus, causing regurgitation, e.g. pulmonary hypertension or cardiomyopathies

- Pathophysiology:
  - Volume-loaded right atrium, leading to dilatation and arrhythmia
  - Late problems: high right atrial pressures resulting in R→L-shunt through a patent foramen ovale PFO
  - Decreased pulmonary blood flow
  - Increased central venous pressure
  - Signs of systemic congestion: peripheral oedema, ascites, etc

- Clinical:
  - Mild to moderate cases: asymptomatic
  - Severe cases: fatigue, signs of right heart failure, oedema
  - Murmur
  - ECG: right atrial enlargement, right ventricular hypertrophy, arrhythmia (atrial fibrillation)
  - Chest x-ray: right heart dilatation, enlargement of superior and inferior vena cava, cardiomegaly in severe cases
  - ECHO: lesion, regurgitant fraction, contractility/ function, additional lesions

Anaesthetic Management
- Avoid bradycardia
- Promote forward flow by decreasing pulmonary vascular resistance: FiO2 >21%, mild hyperventilation
- In case of R→L shunt and cyanosis: secondary erythrocytosis (Chap. 4):
  - coagulation disorder, making them prone to both thrombosis and bleeding
  - risk of acute hyperviscosity syndrome when dehydrated
  - keep well hydrated, minimise fasting times, start IV fluids preoperatively if necessary
- Might need inotropes

Anaesthesia for Repaired Lesion
- Outcome depends on underlying condition
- Either repair or valve replacement
- Arrhythmia may persist
- Valve replacement:
    - anticoagulation management might need stopping or bridging with heparin might be needed, depending on type of valve
    - check local/national guidelines about endocarditis prophylaxis

Additional Reading

1. Anderson J, Grell R, Haines D. Anaesthetic management of a patient with severe pulmonary hypertension, moderate tricuspid regurgitation and moderate right ventricular dysfunction presenting for a dilation and curettage. BMJ Case Rep. 2024;17(2):e257225.
2. Bartlett HL, Atkins DL, Burns TL, Engelkes KJ, Powell SJ, Hills CB, et al. Early outcomes of tricuspid valve replacement in young children. Circulation. 2007;115(3):319–25.

Chapter 32
Total Anomalous Pulmonary Venous Drainage

Abstract In total anomalous pulmonary venous drainage (TAPVD) all pulmonary veins return to the right-sided venous system, either to the right atrium or coronary sinus, or to superior or inferior vena cava or the hepatic veins. This is a L → R shunt lesion with mixing through an intra-atrial communication. Symptoms depend on the intracardiac shunt and degree of obstruction. Anaesthetic management can be challenging and high risk, balancing shunt volume by avoiding decrease in pulmonary vascular resistance with air and mild hypoventilation as well as avoiding an increase in systemic vascular resistance.

Keywords Total anomalous pulmonary venous drainage, TAPVD · Duct-dependent · Shunt lesion, L→R shunt · Mixing lesion, cyanosis · Paediatric cardiac anaesthesia · Congenital cardiac lesion

Supracardiac (~ 50%): pulmonary veins to superior vena cava

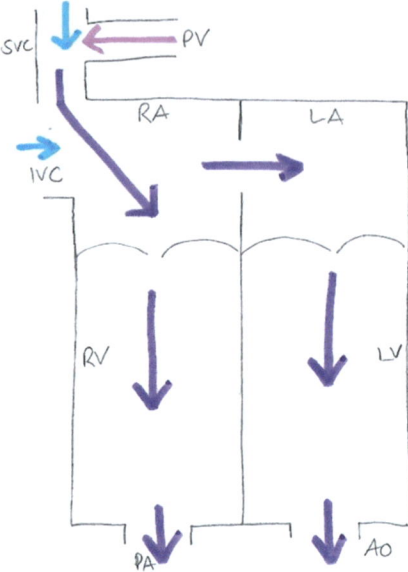

Fig. 32.1 Supracardiac total anomalous pulmonary venous drainage TAPVD (Type I)

Intracardiac (~20%): pulmonary veins to coronary sinus or directly to right atrium

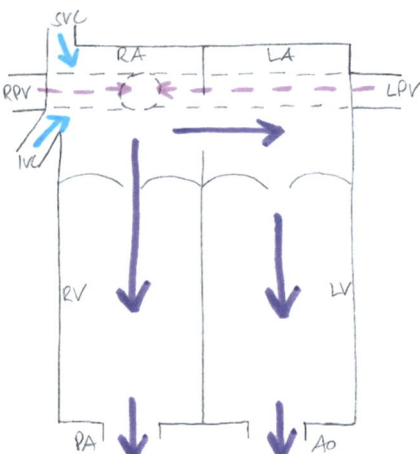

Fig. 32.2 Intracardiac total anomalous pulmonary venous drainage TAPVD (Type II)

Infracardiac (~20%): pulmonary veins to inferior vena cava or hepatic veins

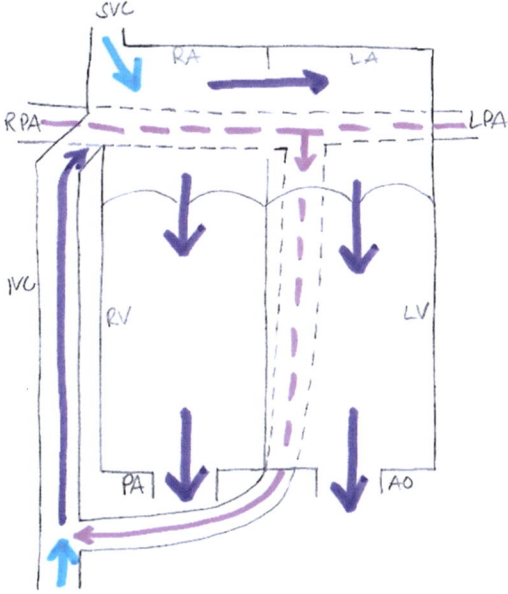

Fig. 32.3 Infracardiac total anomalous pulmonary venous drainage TAPVD (Type III)

Total anomalous pulmonary venous drainage is rare, about 1% of all congenital heart disease. This is a very serious condition which needs surgery early in life. The entire pulmonary venous blood flow returns to the right side of the heart, making it a complete L → R shunt. Survival depends on R → L shunts providing systemic blood flow, either in the heart (atrial and/ or ventricular septal defect, patent foramen ovale) or via a duct. Anaesthetic management is to ensure adequate systemic blood flow for tissue perfusion and oxygenation.

- all pulmonary veins return to the right-sided venous system
- also referred to as anomalous pulmonary venous return (TAPVR) or anomalous pulmonary venous connection (TAPVC)
- isolated defect in about 30%
- 60% associated with additional lesions, most commonly right atrial isomerism (heterotaxy syndrome, Chap. 21)
- complete L → R shunt
- mixing via a R → L shunt via an atrial septal defect ASD or patent foramen ovale PFO: should be unrestricted; if not, either an balloon atrial septostomy (Rashkind procedure, Chap. 41) or emergency surgery needs to be done
- A patent ductus arteriosus might shunt either way or bidirectional
- Types dependent on entry of pulmonary veins (Darling classification):

- Type I: supracardiac (Fig. 32.1)
- Type II: intracardiac (Fig. 32.2)
- Type III: infracardiac (Fig. 32.3)
- Type IV: mixed types: occur in about 10%

- infra-cardiac types are nearly always obstructed and need immediate surgery
- un-obstructed TAPVDs will get obstructed over time and timing of operation will depend on clinical condition and symptoms
- Clinical (no or mild obstruction)

  - mild cyanosis, tachypnoea, tachycardia, failure to thrive
  - might have murmur
  - ECG: right atrial dilatation, right ventricular hypertrophy
  - Chest x-ray: cardiomegaly, pulmonary vein markings
  - Echo: pulmonary veins not to left atrium, atrial septal defect with R → L shunt (unrestrictive?), patent duct and direction of shunt flow, additional lesions?

- with obstruction:

  - cyanosis, congestive heart failure, low cardiac output syndrome LCOS
  - hepatomegaly
  - murmur: gallop rhythm
  - ECG: right atrial dilatation, right ventricular hypertrophy
  - Chest x-ray: pulmonary congestion or oedema
  - Echo: anomalous pulmonary veins to IVC or hepatic veins, atrial septal defect with R → L shunt (unrestrictive?), patent duct and direction of shunt flow, additional lesions?

- **High risk anaesthetic**

Anaesthetic Management

- L → R shunt with overcirculation of pulmonary system

  - Avoid decrease in pulmonary vascular resistance: FiO2 21%, mild hypoventilation
  - Avoid increase in systemic vascular resistance: good anaesthesia and analgesia

- Inotropes might be needed for heart failure
- Right atrial dilatation might cause arrhythmias
- Chronic cyanosis causes secondary erythrocytosis (Chap. 4):

  - coagulation disorder, making them prone to both thrombosis and bleeding
  - risk of acute hyperviscosity syndrome when dehydrated

- keep well hydrated, minimise fasting times, start IV fluids preoperatively if necessary

Anaesthesia for Repaired Lesion

- Full repair depends on associated lesions: good results for biventricular repair
- Atrial arrhythmias might persist due to right atrial pathology and surgery
- Narrowing or stenosis at anastomosis site possible
- Associated abnormalities? Heterotaxy syndrome?
- Residual obstruction or newly developed pulmonary vein disease: common complication after repair, specifically in single ventricle repairs
- If associated lesions make biventricular repair impossible: single ventricle pathway:
    - Initial TAPVD repair +/− further surgery dependent on additional lesions
    - second stage: Glenn shunt (Chap. 38)
    - third stage: total cavo-pulmonary connection TCPC (Chap. 44)

Additional Reading

1. Mohite SN, Divekar VM. Anaesthesia for cervical laminectomy in a patient with total anomalous pulmonary venous connection. Can J Anaesth. 1993;40(6):526–8.
2. Pangasa N, Aravindan A, Subramaniam R, Sirivella P. Total anomalous pulmonary venous connection—anaesthetic management for incidental surgery in a rare presentation. J Anaesthesiol Clin Pharmacol. 2022;38(4):672–3.
3. Ross FJ, Joffe D, Latham GJ. Perioperative and anesthetic considerations in total anomalous pulmonary venous connection. Semin Cardiothorac Vasc Anesth. 2017;21(2):138–44.

Chapter 33
Transposition of the Great Arteries

Abstract Transposition of the great arteries is a duct-dependent, cyanotic lesion, with the pulmonary artery arising from the left ventricle and the aorta from the right ventricle. This lesion needs a patent duct and an intracardiac shunt for mixing for survival.

Balancing the two circulations can be challenging. The shunt via the duct is mainly L → R, but the intracardiac shunts might not be clearly directional. Pulmonary vascular resistance should be kept high with air and mild hypoventilation and an increase in systemic vascular resistance should be avoided.

Keywords Transposition of the great arteries, TGA · Duct-dependent · Shunt lesion · Mixing lesion · Cyanosis · Paediatric cardiac anaesthesia · Congenital cardiac lesion

Transposition of the great arteries make up about 5% of congenital heart disease, with males overrepresented by about 3:1. It is also referred to as D-TGA (dextro-TGA). Symptoms and anaesthetic management depend on the individual details of the lesion and associated defects.

- Pulmonary artery arising from left ventricle, aorta from right ventricle
- Needs shunt/mixing for survival: atrial septal defect, ventricular septal defect, duct (see Fig. 33.1)
- Might have associated cardiac lesions

Fig. 33.1 Transposition of the great arteries TGA: needs shunts for mixing: ASD, VSD, duct

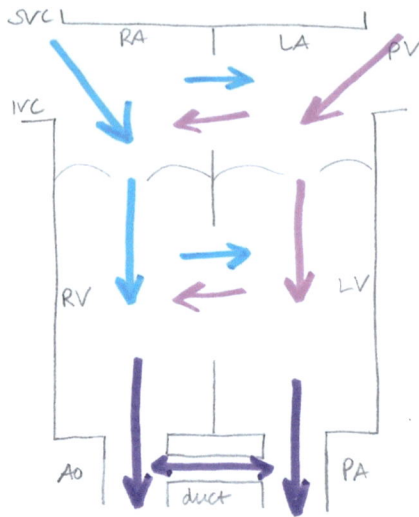

- Clinical:
  - Cyanosis, congestive heart failure
  - Might have murmur from shunts and duct
  - ECG: right ventricular hypertrophy, right atrial dilatation, biventricular hypertrophy in large shunts
  - Chest x-ray: cardiomegaly, egg-shaped heart, narrow superior mediastinum
  - Echo: pulmonary artery from left ventricle, aorta from right ventricle, septal defect size with shunt direction and volume
- **High risk anaesthetic**

Anaesthetic Management

- Mixing and shunt depending on pressures, mostly L → R
  - Avoid decrease in pulmonary vascular resistance: FiO2 21%, mild hypoventilation
  - Avoid increase in systemic vascular resistance: good anaesthesia and analgesia
  - Change pressures as necessary: balance pulmonary and systemic vascular resistance for optimal blood flow and tissue oxygenation
- Duct dependent:
  - On Prostaglandin infusion
  - Keep well hydrated
- With obstructive lesions: avoid tachycardia
- Chronic cyanosis causes secondary erythrocytosis (Chap. 4):
  - coagulation disorder, making them prone to both thrombosis and bleeding
  - risk of acute hyperviscosity syndrome when dehydrated

- keep well hydrated, minimise fasting times, start IV fluids preoperatively if necessary
- Right atrial dilatation can cause arrhythmias

Anaesthesia for Repaired Lesion

- Repair depends on lesion:
    - TGA +/− simple lesion (VSD, PDA, etc): Switch (Chap. 43) + repair of other defects
    - TGA/ VSD/ severe pulmonary stenosis: Rastelli, REV or Nikaidoh (Chap. 43)
    - TGA/ VSD/ subaortic stenosis: Damus-Kaye-Stansel operation (Chap. 37)
- Arterial switch: excellent long term survival
    - Cave obstruction/ stenosis on anastomosis site
    - Coronary blood flow? Coronary obstruction?
- Arrhythmias due to repair of septal defect?
- Other cardiac lesions?

Additional Reading

1. Ribas Ball M, De Miguel Negro M, Galán Menéndez P, Dos Subirà L, Castro Alba MA, Martí Aguasca G. Anesthetic management of pulmonary artery banding in adult patient with single ventricle and uncorrected transposition of the great arteries. Revista Española de Anestesiología y Reanimación (English Edition). 2024 Feb;S2341192924000520.
2. Mathur P, Khare A, Jain N, Verma P, Mathur V. Anesthetic considerations in a child with unrepaired D-transposition of great arteries undergoing noncardiac surgery. Anesth Essays Res. 2015;9(3):440–2.
3. Deal BJ. Late arrhythmias after surgery for transposition of the great arteries. World J Pediatr Cong Heart Surg. 2011;2(1):32–6.
4. Görler H, Ono M, Thies A, Lunkewitz E, Westhoff-Bleck M, Haverich A, Breymann T, Boethig D. Long-term morbidity and quality of life after surgical repair of transposition of the great arteries: atrial versus arterial switch operation. Interact Cardiovasc Thorac Surg. 2011;12(4):569–74.

Chapter 34
Truncus Arteriosus

Abstract Truncus arteriosus is a vascular abnormality with the pulmonary artery and the aorta arising from a common trunk with a single valve, resulting in a large L → R shunt with mixing. Associated lesions are a ventricular septal defect and coronary anomalies. The lesion causes cyanosis, pulmonary hypertension and congestive heart failure. As with every L → R shunt pulmonary vascular resistance should be kept high with air and mild hypoventilation and an increase in systemic vascular resistance should be avoided. Inotropes might be needed for heart failure.

Keywords Ventricular septal defect, truncus arteriosus · VSD, ventricular septal defect · Shunt lesion, vascular lesion · Coronary abnormalities · Cyanosis, mixing lesion · Pulmonary overcirculation · Pulmonary hypertension · Paediatric cardiac anaesthesia · Congenital cardiac lesion

Truncus arteriosus, where both pulmonary artery and aorta come from a single trunk, is a rare lesion, less than 1% of congenital heart disease. This causes mixing of venous and arterial blood and a L → R shunt with overcirculation of the pulmonary system, leading to pulmonary hypertension and congestive heart failure. Anaesthetic management aims to keep the L → R shunt flow to a minimum.

- One single arterial trunk and one single truncal valve
- Ventricular septal defect VSD
- Associated lesions: right aortic arch 30%, DiGeorge 30%
- Coronary abnormalities common
- Several different types, dependent on the origin of pulmonary arteries (see Figs. 34.1, 34.2, 34.3 and 34.4)
- Hemitruncus: only right pulmonary artery RPA arises from aorta (Fig. 34.5)

Fig. 34.1 Truncus arteriosus type 1: main pulmonary artery MPA and aorta Ao arising from a single trunk with a single valve. Occurrence: 60%

Fig. 34.2 Truncus arteriosus type 2: left and right pulmonary artery arise close together from the posterior aorta. Occurrence: 20%

Fig. 34.3 Truncus arteriosus type 3: left and right pulmonary artery arise bilaterally from the aorta. Occurrence: 10%

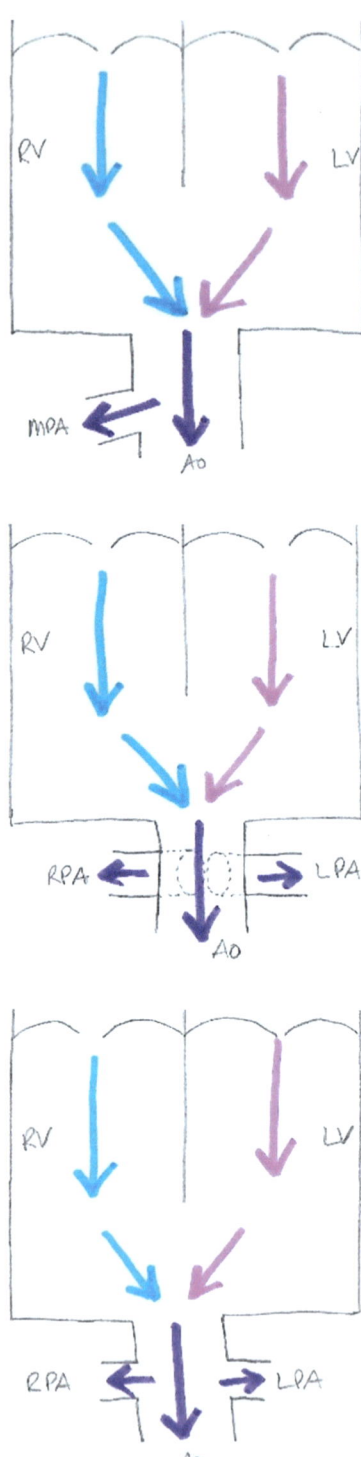

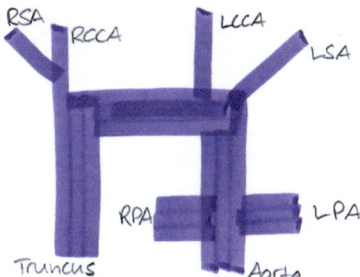

Fig. 34.4 Truncus arteriosus type 4: "Pseudotruncus": severe Tetralogy of Fallot's with pulmonary atresia and aortic collaterals as pulmonary arteries. Occurrence: 10%. *Ao* Aorta, *RPA* right pulmonary artery, *LPA* left pulmonary artery, *RSA* right subclavian artery, *RCCA* right common carotid artery, *LCCA* left common carotid artery, *LSA* left subclavian artery

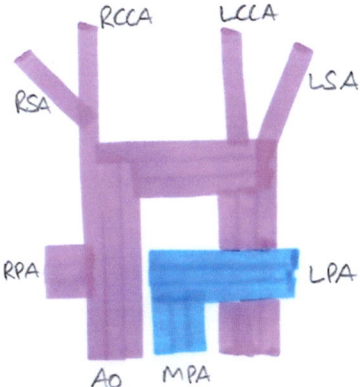

Fig. 34.5 Hemitruncus: right pulmonary artery arising from aorta, left pulmonary artery aring from main pulmonary artery: *Ao* Aorta, *MPA* main pulmonary artery, *RPA* right pulmonary artery, *LPA* left pulmonary artery, *RSA* right subclavian artery, *RCCA* right common carotid artery, *LCCA* left common carotid artery, *LSA* left subclavian artery

- Clinical:
    - Cyanosis, congestive heart failure, pulmonary hypertension
    - Murmur of the ventricular septal defect
    - ECG: biventricular hypertrophy
    - Chest x-ray: cardiomegaly, pulmonary congestion, plethoric lungs
    - Echo: details of pulmonary arteries origin (type of truncus), function/ contractility, ventricular septal defect with size and shunt volume, valve anatomy, coronary anatomy, associated defects?

- **High risk anaesthetic**

Anaesthetic Management
- Severe unrestricted L → R shunt
- Avoid decrease in pulmonary vascular resistance: FiO2 21%, mild hypoventilation
- Avoid increase in systemic vascular resistance: good anaesthetic and analgesia
- Coronary abnormalities: keep adequate perfusion pressure
- Inotropes might be needed for heart failure
- Chronic cyanosis causes secondary erythrocytosis (Chap. 4):
    - coagulation disorder, making them prone to both thrombosis and bleeding
    - risk of acute hyperviscosity syndrome when dehydrated
- keep well hydrated, minimise fasting times, start IV fluids preoperatively if necessary

Anaesthesia for Repaired Lesion
- Usually full repair possible: truncal valve = aortic valve and conduit from right ventricle to pulmonary artery (RV-PA) to provide pulmonary blood flow
- Arrhythmias due to ventriculotomy/ ventricular septal defect closure
- Check Echo for:
    - Anastomosis unobstructed? Stenosis, narrowing, calcification?
    - Right ventricular to pulmonary artery conduit: blood flow unobstructed?
    - Coronary anatomy and blood flow?
    - Contractility/ function

Additional Reading

1. Fischer MU, Priebe HJ. Anaesthetic management for hip arthroplasty in a 46-yr-old patient with uncorrected truncus arteriosus type IV. Br J Anaesth. 2006;97(3):329–32.
2. Parikh R, Eisses M, Latham GJ, Joffe DC, Ross FJ. Perioperative and Anesthetic Considerations in Truncus Arteriosus. Semin Cardiothorac Vasc Anesth. 2018;22(3):285–93.
3. Díaz-Fosado LA, Sarmiento L, Velazquez-Martínez T. Anesthetic management of a schoolboy with uncorrected truncus arteriosus type I, and severe pulmonary hypertension undergoing repair of congenital dislocation of the knee. Case report☆. Colomb J Anesthesiol. 2016;44(3):259.
4. Bosman M. Truncus arteriosus: perioperative management. Egypt J Cardiothorac Anesth. 2012;6(1):1–6.

Chapter 35
Ventricular Septal Defect

Abstract Ventricular septal defects are L → R shunt lesions, with symptoms depending on the size of the shunt. Small defects are usually asymptomatic, large or long-standing shunts can cause heart failure, pulmonary hypertension and shunt reversal (Eisenmenger's syndrome, resulting in a R → L shunt with cyanosis).

As with all L → R shunts, a decrease in pulmonary vascular resistance should be avoided with air and mild hypoventilation. An increase in systemic vascular resistance should be avoided as well. Inotropes might be needed in heart failure.

Keywords VSD · Ventricular septal defect · Shunt lesion · Shunt reversal · Eisenmenger's syndrome · Pulmonary overcirculation · Pulmonary hypertension · Paediatric cardiac anaesthesia · Congenital cardiac lesion

Ventricular septal defects are the most common congenital heart defect with about 15–20% of all congenital heart lesions. This lesion causes a L → R shunt with overcirculation of the pulmonary system, over time leading to pulmonary hypertension and right heart failure. Symptoms depend on size and position of defect. Anaesthetic management aims to keep shunt volume as low as possible.

- Types:
    - Perimembranous: most common, does not close spontaneously
    - Muscular → might close spontaneously
- Might involve valvular tissue (prolapse into defect)
- Might be several defects
- Associated with other cardiac lesions

- Clinical:
  - Symptoms dependent on size of defect
- Small: asymptomatic, maybe breathless on exertion (Fig. 35.1)
- Large: congestive heart failure, pulmonary hypertension PHT (Fig. 35.2)
- Heart murmur from the shunt or tricuspid regurgitation
- ECG: normal; in large lesions: left ventricular hypertrophy, right ventricular hypertrophy
- Chest x-ray: normal; in large defects: cardiomegaly, plethoric lungs
- Echo: small defect: Left heart volume load and dilatation; large defect: biventricular hypertrophy, dilatation, dysfunction, pulmonary hypertension with tricuspid regurgitation

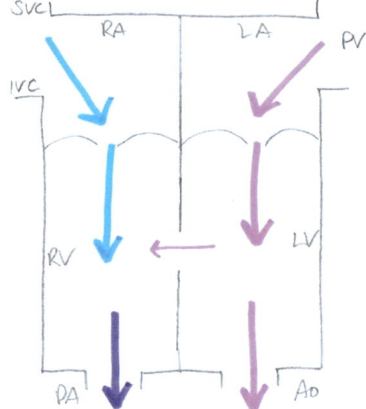

Fig. 35.1 Ventricular septal defect, small

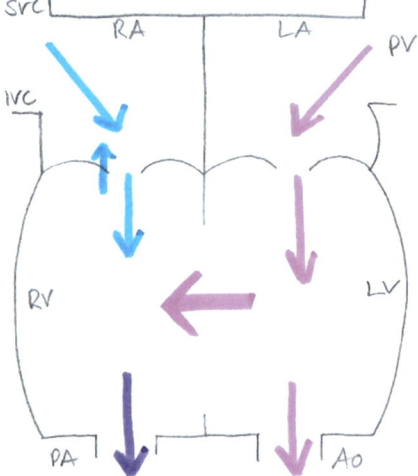

Fig. 35.2 Ventricular septal defect, large, with volume load, dilatation and tricuspid regurgitation

Anaesthetic Management
- L → R shunt
  - Avoid decrease in pulmonary vascular resistance: FiO2 21%, mild hypoventilation
  - Avoid increase in systemic vascular resistance: good anaesthesia and analgesia
- Problems with late stages:
  - Volume load and dilatation of ventricles: impaired function: might need inotropes
  - Stroke volume can no longer be improved by preload
  - Over time the increase in pulmonary blood flow will cause compensatory pulmonary vasoconstriction, followed by hyperplasia and irreversible fibrosis of the pulmonary vascular bed, causing pulmonary hypertension, tricuspid regurgitation and right ventricular failure
- In pulmonary hypertension FiO2 > 21% might be beneficial
- Careful fluid management

Anaesthesia for Repaired Lesion
- Either surgical repair or device closure
- Usually no precautions necessary
- Might have arrhythmias due to surgical manipulation of conduction tissue or pacemaker
- Echo:
  - Residual shunt?
  - Persistent impaired function?
  - Persistent pulmonary hypertension?
  - Associated lesions?

Additional Reading

1. Yen P. ASD and VSD Flow Dynamics and Anesthetic Management. Anesth Prog. 2015;62(3):125–30.
2. Pathak P, Das S, Gupta S, Hasija S, Choudhury A, Gharde P, et al. Effect of change in tidal volume on left to right shunt across ventricular septal defect in children—A pilot study. Ann Pediatr Card. 2021;14(3):350–5.
3. Li P, Zeng J, Wei W, Lin J. The effects of ventilation on left-to-right shunt and regional cerebral oxygen saturation: a self-controlled trial. BMC Anesthesiol. 2019 Dec;19(1):178.
4. Shivaramu BT, Shashank MR, Prajwal PH. Anaesthetic management of an adult patient with uncorrected ventricular septal defect posted for obstructed inguinal hernia repair. J Evol Med Dent Sci. 2015;4(66):11574–80.

Part II
Operations

Chapter 36
Central Shunts

Abstract Central shunts are artificial connections between the systemic and the pulmonary circulation to provide pulmonary blood flow. This is usually part of a complex repair or a palliative operation for patients in whom complete repair is not possible. There are several different types, depending on the anatomy and blood vessels involved.

In a L → R shunt a decrease in pulmonary vascular resistance and an increase in systemic vascular resistance should be avoided. Anaesthetic management might be different if shunt is part of a complex repair.

Keywords Central shunt, BT shunt, Blalock-Taussig shunt · Laks shunt, melbourne shunt, waterston shunt · Potts shunt, cooley shunt · Paediatric cardiac anaesthesia · Congenital cardiac lesion

A central shunt is a connection between the systemic and pulmonary circulation, providing blood flow between the two. It is part of a complex repair or a palliative operation for patients in whom complete repair is not possible. In the majority of cases the shunt flow is from left to right, meaning that the systemic circulation provides blood flow to the pulmonary system. The shunt can be direct, i.e. an anastomosis of the aorta and the pulmonary artery, or a synthetic tubular graft, usually made from goretex. Several of these shunts are no longer done due to high complication rates and improved surgical techniques for the repair of complex cases, but might still be seen in older patients.

Anaesthetic management is that of a shunt lesion, with consideration for the underlying complex heart defect in case of a palliative operation.

1. Blalock-Taussig shunt BTS (also referred to as Blalock-Taussig-Thomas shunt):
   - Classic BT shunt (Fig. 36.1): end to side anastomosis of subclavian artery to pulmonary artery, on the opposite side to the aortic arch to avoid kinking of subclavian artery
   - Modified BT shunt (Fig. 36.2): goretex tube from subclavian artery to pulmonary artery
2. Laks shunt (Fig. 36.3): goretex tube from aorta to main pulmonary artery
3. Melbourne shunt: end to side anastomosis of main pulmonary artery to aorta
4. Potts shunt (Fig. 36.4): side to side anastomosis of descending aorta to pulmonary artery: palliative procedure before complete repair was possible, now obsolete
5. Reverse Potts shunt: side to side anastomosis of descending aorta to pulmonary artery in patients with supra-systemic pulmonary hypertension: R → L shunt
6. Waterston shunt (Fig. 36.5): side to side anastomosis of ascending aorta to pulmonary artery (obsolete)
   - variation: Cooley shunt: intrapericardial shunt from ascending aorta to right pulmonary artery (obsolete)

Anaesthetic Management
- Check Echo for shunt position, size and shunt volume, details of underlying lesion, contractility/function
- L → R shunt
  - Keep pulmonary vascular resistance high with FiO2 21% and mild hypoventilation
  - Keep systemic vascular resistance normal/low, avoid an increase in systemic blood pressure: good anaesthesia and analgesia
  - Adjust pressures if needed
- Pulmonary blood flow dependent on pressure difference between pulmonary and systemic pressures and shunt size
- Conduits/grafts can calcify, get stenotic or kink → limited pulmonary blood flow

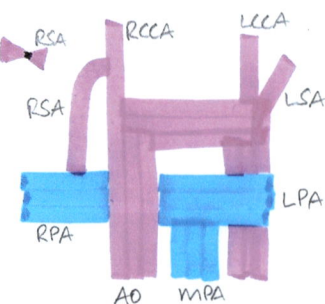

Fig. 36.1 Classic Blalock-Taussig shunt: attachment of subclavian artery to the pulmonary artery and ligation of the distal end of the subclavian artery. *Ao* Aorta, *MPA* main pulmonary artery, *LPA* left pulmonary artery, *RPA* right pulmonary artery, *RSA* right subclavian artery, *RCCA* right common carotid artery, *LCCA* left common carotid artery, *LSA* left subclavian artery

36 Central Shunts

Fig. 36.2 Modified Blalock-Taussig shunt: goretex tube from subclavian artery to the pulmonary artery. *Ao* Aorta, *PA* pulmonary artery, *RSA* right subclavian artery, *RCCA* right common carotid artery, *LCCA* left common carotid artery, *LSA* left subclavian artery

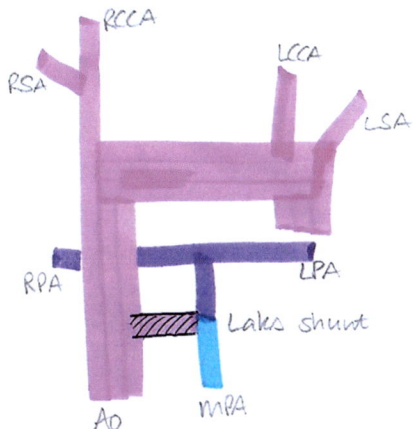

Fig. 36.3 Laks shunt: goretex tube connection between aorta and main pulmonary artery to promote blood flow into pulmonary arteries for growth. *Ao* Aorta, *MPA* main pulmonary artery, *RPA* right pulmonary artery, *LPA* left pulmonary artery, *RSA* right subclavian artery, *RCCA* right common carotid artery, *LCCA* left common carotid artery, *LSA* left subclavian artery

Fig. 36.4 Potts shunt: connection between descending aorta Ao and left pulmonary artery

Fig. 36.5 Waterston shunt: connection between ascending aorta Ao and right pulmonary artery

Additional Reading

1. Kiran U, Aggarwal S, Choudhary A, Uma B, Kapoor PM. The blalock and taussig shunt revisited. Ann Card Anaesth. 2017;20(3):323–30. https://doi.org/10.4103/aca.ACA_80_17.
2. Gupta B, Gupta A, Agarwal M, Gupta L. Glenn shunt: anaesthetic concerns for a non cardiac surgery. North J ISA. 2017;2:36–42.

Chapter 37
Damus-Kaye-Stansel Operation

Abstract Damus-Kaye-Stansel (DKS) is an operation used for relief of systemic outflow tract obstruction, usually in a single ventricle circulation. The unobstructed pulmonary outflow is connected to the aorta and pulmonary blood flow is provided by a modified BT shunt or an right ventricle to pulmonary artery conduit.

Anaesthetically the pulmonary vascular resistance should be kept high with air and mild hypoventilation, but in case of an obstructed/ stenosed or kinked conduit, oxygen and mild hyperventilation can be used to improve pulmonary blood flow. Inotropes might be needed for impaired function.

Keywords Damus-kaye-stansel, DKS · RV-PA conduit · BT shunt · Single ventricle · Mixing lesion · Cyanosis · Paediatric cardiac anaesthesia · Congenital cardiac lesion

A Damus-Kaye-Stansel operation is used for relief of systemic outflow tract obstruction, usually in a single ventricle circulation. This surgery commits both main arteries to the systemic circulation and provides pulmonary blood flow via a right ventricle to pulmonary artery shunt (Sano shunt) or a Blalock-Taussig shunt (Chap. 36).

- the operation consists of (see Fig. 37.1):
    - Closure of main pulmonary artery
    - Connection of proximal end of pulmonary artery to aorta
    - Pulmonary blood flow via Blalock-Taussig shunt, classic or modified, (Chap. 36) or Sano shunt (right ventricle to pulmonary artery conduit)
- Mixing lesion
- Saturations 75–85%

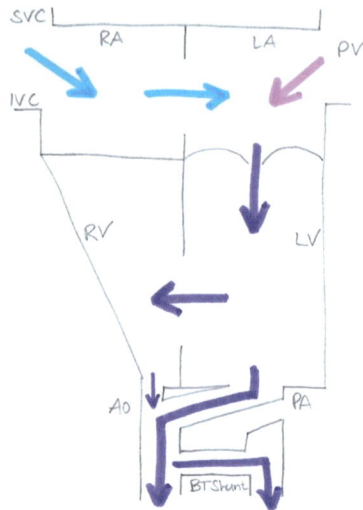

Fig. 37.1 Damus-Kaye-Stansel operation for tricuspid atresia TA with transposition of the great arteries TGA. A classic Blalock-Taussig shunt BTS (connection of subclavian artery to pulmonary artery) provides pulmonary blood flow

Anaesthetic Management
- Keep fasting times to a minimum, start IV fluids pre-operatively, if necessary
- Maintain preload, keep well hydrated
- might have impaired ventricular function due to underlying condition → inotropes
- Chronic cyanosis causes secondary erythrocytosis (Chap. 4):
    - coagulation disorder, making them prone to both thrombosis and bleeding
    - risk of acute hyperviscosity syndrome when dehydrated
- modified Blalock-Taussig shunt or Sano shunt might be narrowed/ stenosed or kinked
    - reduced pulmonary blood flow
        → decrease pulmonary vascular resistance with FiO2 > 21% and mild hyperventilation
    - obstructed Sano shunt causing increased right ventricular pressures, leading to ventricular impairment, dysfunction and failure
        → inotropic support

Additional Reading

1. Alsoufi B. Management of the single ventricle and potentially obstructive systemic ventricular outflow tract. J Saudi Heart Assoc. 2013 Jul;25(3):191–202.

Chapter 38
Glenn Shunt

Abstract A Glenn shunt is a connection of the superior vena cava to the pulmonary artery to provide pulmonary blood flow. This operation is the second step in the pathway to a single ventricle circulation. Patients are usually cyanotic with polycythaemia, which can cause hyperviscosity syndrome when dehydrated and coagulation problems.

Anaesthetic management is determined by the underlying condition. For good flow through the Glenn keep well hydrated to maintain central venous pressure and carefully decrease the pulmonary vascular resistance with oxygen and mild hyperventilation as needed.

Keywords Glenn shunt · Cyanosis · Single ventricle · Hemifontan · Shunt · Mixing lesion · Paediatric cardiac anaesthesia · Congenital cardiac lesion

A Glenn shunt is a veno-arterial shunt, connecting the superior vena cava to the pulmonary artery, redirecting the systemic venous return from the upper body to the pulmonary circulation. Shunt volume and blood flow is dependent on a pressure difference between the central venous pressures and the pulmonary vascular resistance. A variation of this shunt is the Hemifontan, in which the superior vena cava is connected to the pulmonary artery via the right atrium. Since the inferior vena cava with the venous return from the lower body still drains into the right heart, there will still be a pulsatile flow into the pulmonary circulation.

- Glenn shunt: connection of the superior vena cava to the pulmonary artery (Fig. 38.1)
    - usually bidirectional, so providing blood flow to both lungs
    - can be bilateral in case of bilateral superior vena cava

Fig. 38.1 Glenn shunt: connection of superior vena cava SVC to the pulmonary artery PA

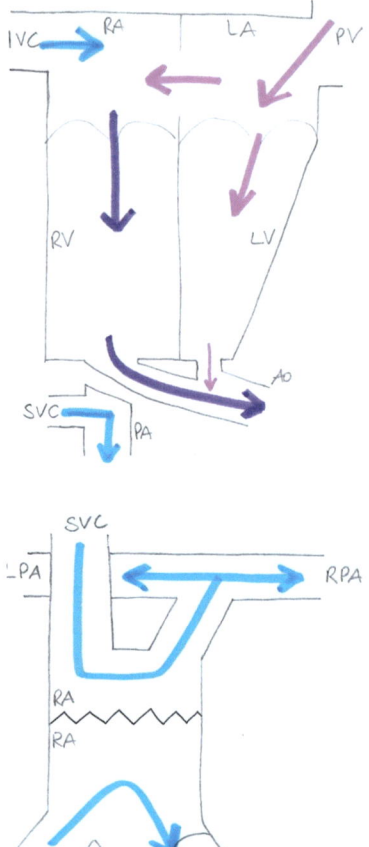

Fig. 38.2 Hemifontan: a membrane (visualised as a zig-zag line) divides the right atrium RA to direct blood from the superior vena cava SVC to the main pulmonary artery, while blood from the inferior vena cava IVC flows into the right ventricle RV

- Hemifontan (Fig. 38.2): superior part of the right atrial appendage is connected to the main pulmonary artery and an intra-atrial baffle re-routes the blood from the superior vena cava to the pulmonary artery
- usually done at 3–6 months of age, once the pulmonary vascular resistance PVR has decreased from birth levels
- second stage operation for single ventricle circulation, e.g. Hypoplastic Left Heart Syndrome HLHS, Tricuspid Atresia, Ebstein's Anomaly, Double Outlet Right Ventricle DORV, Double Inlet Left Ventricle DILV, etc.
- Mixing lesion
- Saturations 75–85%

Anaesthetic Management
- Premedication is advisable if underlying condition is dependent on low pulmonary vascular resistance
- Chronic cyanosis causes secondary erythrocytosis (Chap. 4):
  - coagulation disorder, making them prone to both thrombosis and bleeding
  - risk of acute hyperviscosity syndrome when dehydrated
- Keep hydrated, minimise starvation times, start IV fluids preoperatively if necessary
- Check clotting and anti-coagulation medication
- Mixing lesion: balance PVR/ SVR for optimal saturations and oxygen delivery
- Ideally keep saturation and pressures at pre-op values
- Keep pulmonary vascular resistance at pre-op level:
  - FiO_2 21%
  - Mild hypoventilation
- Keep systemic vascular resistance at pre-op level:
  - Good anaesthesia and analgesia
  - Fluids, vasopressors if needed
- If saturations drop, decrease pulmonary vascular resistance by increasing FiO_2 and ventilation
- Up to 25% of patients develop pulmonary arterio-venous malformations, causing a R → L shunt, resulting in a decrease of saturations

Additional Reading

1. Christensen RE, Gholami AS, Reynolds PI, Malviya S. Anaesthetic management and outcomes after noncardiac surgery in patients with hypoplastic left heart syndrome: a retrospective review. Eur J Anaesthesiol. 2012;29(9):425–30.
2. Gupta B, Gupta A, Agarwal M, Gupta L. Glenn shunt: anaesthetic concerns for a non cardiac surgery. Northern J ISA. 2017;2:36–42.

Chapter 39
Hybrid Procedure for Hypoplastic Left Heart Syndrome

Abstract The hybrid procedure is an alternative operation to the Norwood operation for hypoplastic left heart syndrome. It consists of pulmonary artery bands, atrial septostomy and a ductal stent. Patients are cyanotic with polycythaemia, which can cause hyperviscosity syndrome when dehydrated and coagulation problems. Anaesthesia is challenging with a R → L shunt through the duct. Saturations should be kept at 75–85%. Pulmonary blood flow is limited by the pulmonary bands, so increasing oxygen and ventilation might not result in increased saturations.

Keywords Hybrid operation, hybrid procedure · HLHS, hypoplastic left heart syndrome · Single ventricle · Pulmonary artery bands · Atrial septostomy · Ductal stent · Mixing · Cyanosis · Polycythaemia · Hyperviscosity syndrome · Paediatric cardiac anaesthesia · Congenital cardiac lesion

A Hybrid procedure is an alternative procedure to the Norwood operation for Hypoplastic Left Heart Syndrome (Fig. 39.1). It is usually done in neonates with low birth weight, who are too small or fragile for a Norwood operation.

- Consists of (Fig. 39.2):
    - Bilateral pulmonary artery bands to limit pulmonary blood flow and prevent pulmonary hypertension
    - Ductal stent to provide adequate systemic perfusion
    - Atrial septostomy
- Advantages:
    - no cardiopulmonary bypass necessary
    - lower mortality (~2.5%) than Norwood operation (5–20%)

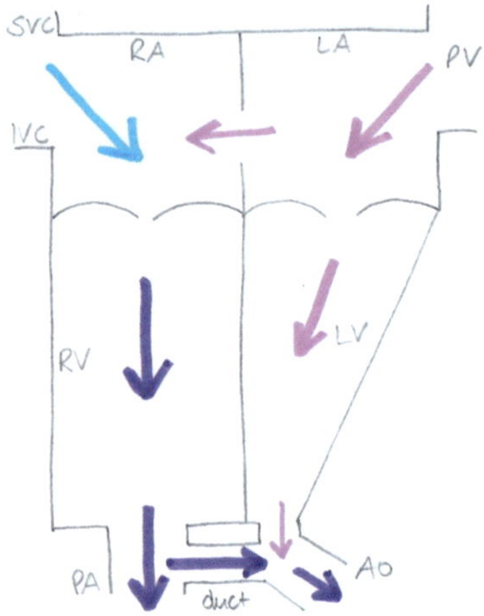

Fig. 39.1 Hypoplastic left heart anatomy and blood flow

Fig. 39.2 Hybrid procedure: atrial septostomy, pulmonary artery bands and ductal stent

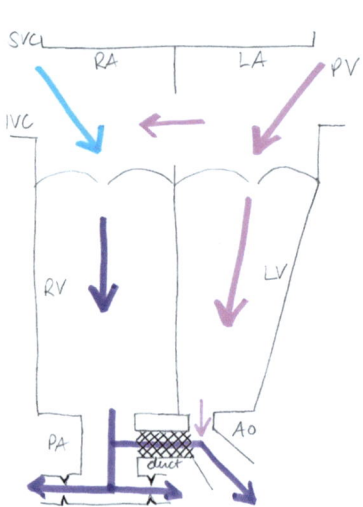

- As the baby grows, the pulmonary artery bands will get restrictive and limit the pulmonary blood flow
- Followed by Norwood and Glenn at ~3–6 months of age
- Mixing lesion
- Cyanotic, saturation 75–85%
- Chronic cyanosis causes secondary erythrocytosis (Chap. 4)

    – coagulation disorder, making them prone to both thrombosis and bleeding
    – risk of acute hyperviscosity syndrome when dehydrated

- **High risk anaesthesia**

Anaesthetic Management
- Keep well hydrated, minimise fasting time, start IV fluids preoperatively if necessary
- Echo:
    - flow through the atrial septum and the ductal stent should be unrestricted
    - contractility/ function?
    - Flow through pulmonary artery bands: limited? Obstructive?
    - Additional lesions?
- Limited pulmonary blood flow due to pulmonary artery bands → increasing oxygen and ventilation might not result in increased saturations
- Maintain preload
- Might need inotropes for impaired function

Additional Reading

1. Naguib AN, Winch P, Schwartz L, Isaacs J, Rodeman R, Cheatham JP, et al. Anesthetic management of the hybrid stage 1 procedure for hypoplastic left heart syndrome (HLHS). Pediatr Anesth. 2010;20(1):38–46.
2. Lok T, Winch P, Naguib A. Perioperative management of a child with hypoplastic left heart syndrome following the hybrid stage I procedure presenting for laparoscopic gastrostomy tube placement. Pediatr Anesth Crit Care J. 2017;5(1):24–30.
3. Christensen RE, Gholami AS, Reynolds PI, Malviya S. Anaesthetic management and outcomes after noncardiac surgery in patients with hypoplastic left heart syndrome: a retrospective review. Eur J Anaesthesiol. 2012;29(9):425–30.

Chapter 40
Norwood Operation

Abstract The Norwood operation is the first operation for hypoplastic left heart syndrome in the pathway to a single ventricle circulation. It consists of the connection of pulmonary artery (proximal end) to aorta, closure of the distal end of the pulmonary artery, enlargement of the hypoplastic aorta with a patch and an atrial septectomy. Pulmonary blood flow is provided via a Blalock-Taussig shunt or Sano shunt.

This is a mixing lesion, both pulmonary vascular resistance and systemic vascular resistance should be kept at pre-operative levels.

Keywords Norwood operation, HLHS · Atrial septostomy, blalock-taussig shunt · Hypoplastic left heart syndrome, BT shunt · Single ventricle, cyanosis, mixing lesion · Sano shunt, RV-PA conduit · Paediatric cardiac anaesthesia · Congenital cardiac lesion

The Norwood operation is surgery for hypoplastic left heart and hypoplastic aortic arch (Fig. 40.1) which is usually associated with small or atretic mitral and/or aortic valves.

- Consists of:
    - Main pulmonary artery is ligated (distal end)
    - Connection of pulmonary artery (proximal end) to aorta
    - Enlargement of aorta with patch
    - Pulmonary blood flow via Blalock-Taussig shunt (Fig. 40.2; Chap. 36) or Sano shunt (Right ventricle to pulmonary artery conduit, Fig. 40.3)
    - Excision of atrial septum to create an unrestricted atrial L → R shunt
- Interstage mortality to Glenn shunt (Chap. 38) at ~3–6 months of age is up to 30%
- Saturations above 75% are acceptable, low to mid 80s is optimal, >90% is over-circulating the lungs

Fig. 40.1 Hypoplastic left heart anatomy and blood flow

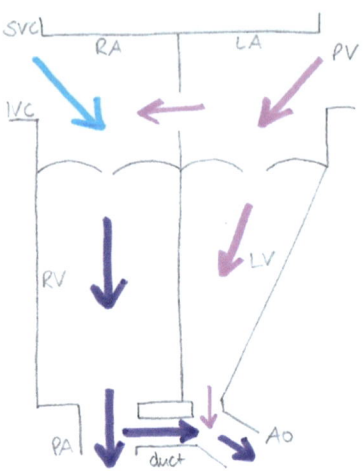

Fig. 40.2 Norwood procedure with Blalock-Taussig shunt BTS: the shaded area on the aorta signifies the aortic enlargement patch. *LSCA* left subclavian artery

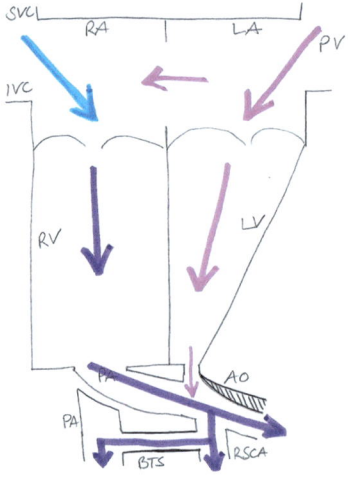

Fig. 40.3 Norwood with Sano shunt: a right ventricle to pulmonary artery (RV-PA) conduit to provide pulmonary blood flow. The shaded area on the aorta signifies the aortic enlargement patch

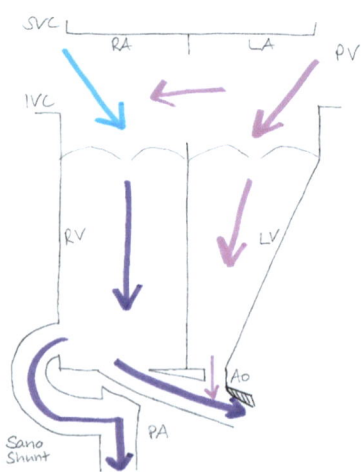

- **High risk anaesthesia**

Anaesthetic Management
- Chronic cyanosis causes secondary erythrocytosis (Chap. 4):
  - coagulation disorder, making them prone to both thrombosis and bleeding
  - risk of acute hyperviscosity syndrome when dehydrated
- keep well hydrated, minimise fasting times, start IV fluids preoperatively if necessary
- L → R shunt at atrial level → needs to be unrestricted
- with BT shunt: L → R shunt:
  - avoid decrease in pulmonary vascular resistance: FiO2 21% and mild hypoventilation
  - avoid increase in systemic vascular resistance: good anaesthetic, vasodilators
- Mixing lesion: keep pulmonary and systemic blood pressure at pre-op level
  - Pulmonary vascular resistance: FiO2 21% and mild hypoventilation
  - Systemic vascular resistance: fluids, good anaesthetic, vasopressors or vasodilators as needed
  - Change pressures as needed, balance pulmonary and systemic blood flow to ensure optimal perfusion pressures and tissue oxygenation
- If saturations <75%:
  - Mild increase in FiO2, normo-ventilation to decrease pulmonary vascular resistance to increase blood flow to the pulmonary circulation
  - In BT shunt: either decrease pulmonary vascular resistance or increase systemic vascular resistance to increase blood flow into pulmonary circulation
- Might need inotropic support: right ventricle supports both circulations
- Careful fluid management throughout

Additional Reading

1. Christensen RE, Gholami AS, Reynolds PI, Malviya S. Anaesthetic management and outcomes after noncardiac surgery in patients with hypoplastic left heart syndrome: a retrospective review. Eur J Anaesthesiol. 2012;29(9):425–30.
2. Mariano ER, Boltz MG, Albanese CT, Abrajano CT, Ramamoorthy C. Anesthetic management of infants with palliated hypoplastic left heart syndrome undergoing laparoscopic nissen fundoplication. Anesth Analg. 2005;100(6):1631–3.

Chapter 41
Palliative Procedures

Abstract If cardiac lesions cannot be repaired, they are palliated with a variety of operations. The three most common are ductal stent, pulmonary artery band and balloon atrial septostomy (Rashkind procedure). The anaesthetic management is determined by both the underlying lesion and the palliative procedure. In a pulmonary artery band the pulmonary blood flow is limited by the tightness of the band and might limit cardiac output. A balloon atrial septostomy is done in lesions where an unrestrictive L → R shunt and mixing is needed for survival. Flow through a ductal stent can be L → R or R → L, depending on the underlying obstruction.

Keywords Palliative operation, palliative procedure · Ductal stent, pulmonary artery band · Balloon atrial septostomy, Rashkind procedure · Shunt lesion · Duct-dependent lesion · Paediatric cardiac anaesthesia · Congenital cardiac lesion

Palliative procedures do not repair a lesion, but serve as a bridge to full repair. They are usually done in babies too small for full repair or not suitable for full repair.

41.1 Pulmonary Artery Band (Fig. 41.1)

- limits pulmonary blood flow
  - protects pulmonary vascular bed from excessive blood flow
  - prevents pulmonary hypertension
  - can be used to "train" left ventricle prior to transposition repair
- Can be done for main pulmonary artery or branch pulmonary arteries

Fig. 41.1 Pulmonary artery band in double outlet right ventricle DORV to restrict pulmonary blood flow

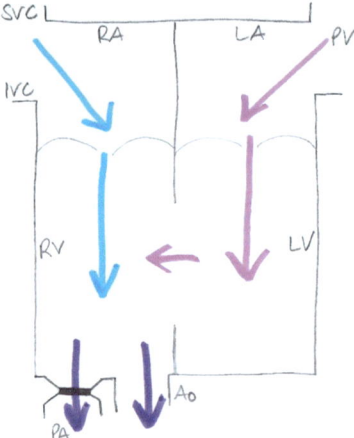

- As heart/ blood vessels grow, pulmonary artery band will become more restrictive over time: gradient will increase and ventricular pressure will increase → ventricular hypertrophy

Anaesthetic Management
- Oxygen (FiO2 > 21%) will not result in overshunting the pulmonary vasculature, but might cause pulmonary blood pooling by vasodilatation
- increase in ventricular pressures due to artificial pulmonary stenosis might result in impaired function → might need inotropes
- maintain preload
- underlying lesion?

41.2 Balloon Atrial Septostomy (Rashkind Procedure, Fig. 41.2)

- creation of an unrestrictive atrial septal defect
- increases mixing at atrial level for lesions depending on unrestrictive mixing, like tricuspid atresia, transposition of the great arteries, total anomalous pulmonary venous drainage, etc.

Anaesthetic Management
- usually no problems from the unrestrictive atrial septum itself
- challenging anaesthetic due to the underlying complex cyanotic lesion
- might need inotropes for impaired function
- Chronic cyanosis causes secondary erythrocytosis (Chap. 4)

  - coagulation disorder, making them prone to both thrombosis and bleeding
  - risk of acute hyperviscosity syndrome when dehydrated

- keep well hydrated, minimise fasting times, start IV fluids preoperatively if necessary

Fig. 41.2 Restrictive atrial septal defect in tricuspid atresia: need for balloon atrial septectomy

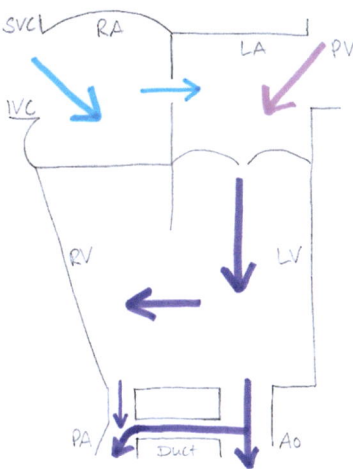

41.3 Ductal Stent (Fig. 41.3)

- insertion of stent into ductus arteriosus for duct-dependent lesions, such as pulmonary atresia, tricuspid atresia, severe Ebstein's, etc.
- Creating a permanent shunt without the need for continuous prostaglandin infusions
- Shunt flow can be L → R to provide pulmonary blood flow or R → L to provide systemic blood flow.

Anaesthetic Management
- Regardless of direction of shunt flow: pulmonary pressures are usually lower than systemic pressures, so flow will be predominantly into pulmonary circulation
- Avoid decrease in pulmonary vascular resistance: FiO2 21%, mild hypoventilation
- Avoid increase in systemic vascular resistance: good anaesthesia and analgesia
- Maintain preload
- Might need inotropes due to impaired function
- Challenging anaesthesia due to underlying complex heart disease

Fig. 41.3 Ductal stent (represented by black cross hatching) in pulmonary atresia with intact ventricular septum PA/ IVS

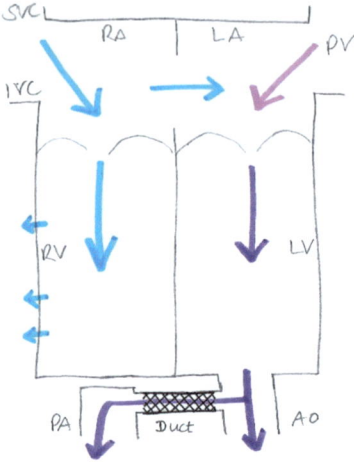

Additional Reading

1. Ribas Ball M, De Miguel Negro M, Galán Menéndez P, Dos Subirà L, Castro Alba MA, Martí Aguasca G. Anesthetic management of pulmonary artery banding in adult patient with single ventricle and uncorrected transposition of the great arteries. Revista Española de Anestesiología y Reanimación (English Edition). 2024 Feb;S2341192924000520.

Chapter 42
Aortic Valve Operations

Abstract Dysplastic aortic valves or aortic roots need to be replaced or repaired. The Ross and Konno operations are the most common. In a Ross operation the aortic valve is replaced by the pulmonary valve, with a homograft as the new pulmonary valve. The Konno operation is an enlargement of aortic valve annulus and aortic root by extension of the ventricular septum with a patch, usually followed by a aortic valve replacement.

Anaesthetic management depends on associated lesions and the quality of the repair.

Keywords Aortic valve · Ross operation · Konno operation · Valvular lesions · Ross-konno · Arryhthmia · Paediatric cardiac anaesthesia · Congenital cardiac lesion

42.1 Ross Operation

A Ross operation is a replacement of native aortic valve with the native pulmonary valve and placement of homograft, bioprosthetic or mechanical valve for the pulmonary valve (Figs. 42.1 and 42.2).

Anaesthetic Management
- Coronary re-implantation → check Echo for good function, adequate flow and stenosis/obstruction
- Might be part of a complex repair → check Echo for residual lesions and myocardial function
- Maintain preload
- Complex lesions/repair: might need inotropes

Fig. 42.1 Ross operation: the native pulmonary valve gets excised and placed as the new aortic valve (neo-aortic valve)

Fig. 42.2 the pulmonary valve is replaced by a homograft valve

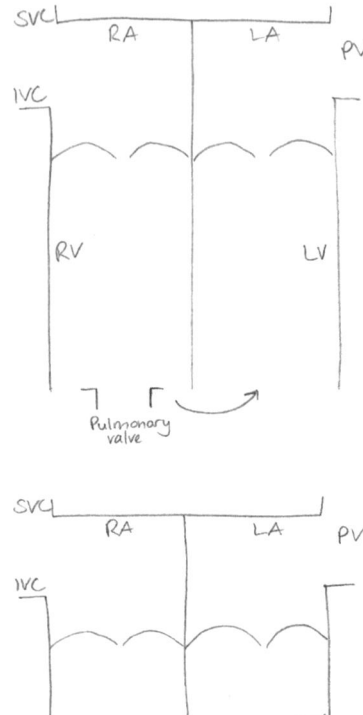

- Anti-coagulation medication for mechanical valve: depending on surgery might need stopping or bridging with heparin might be needed

42.2 Konno Operation

A Konno operation is an aortic root replacement with ventricular septal extension (Fig. 42.3), sometimes followed by replacement of the native aortic valve with either the pulmonary valve (Ross operation), a homograft, bioprosthetic or mechanical valve.

Fig. 42.3 Konno operation: enlargement of aortic valve annulus and aortic root by extension of the ventricular septum with a patch, represented here by the zig-zag line

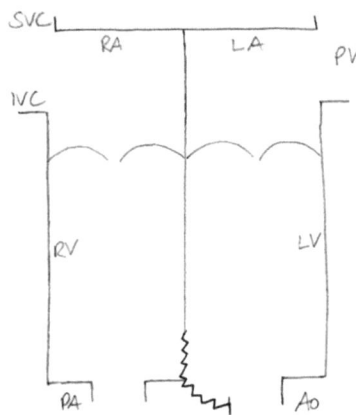

Anaesthetic Management
- might have arrhythmias due to manipulation of pacing tissue in the ventricular septum
- might have pacemaker
- maintain preload
- might have been part of a complex repair for cyanotic heart disease → Echo about residual lesion, function, coronaries, etc.
- might need inotropes for impaired function
- anti-coagulation medication for mechanical valve: depending on surgery might need stopping or bridging with heparin might be needed

Additional Reading

1. Schlein J, Ebner BE, Geiger R, Simon P, Wollenek G, Moritz A, et al. Long-term outcomes after the paediatric Ross and Ross-Konno procedures. Interact Cardiovasc Thorac Surg. 2021;33(3):455–61.
2. Watabe A, Saito H, Harasawa K, Morimoto Y. Anesthetic management for severe aortic regurgitation in an infant repaired by Ross procedure. J Anesth. 2009;23(2):270–4.
3. Zaidi M, Premkumar G, Naqvi R, Khashkhusha A, Aslam Z, Ali A, et al. Aortic valve surgery: management and outcomes in the paediatric population. Eur J Pediatr. 2021;180(10):3129–39.

Chapter 43
Switch Operations

Abstract There are several options for repair of transposition of the great arteries. Individual anatomy determines the decision for atrial, ventricular or arterial switch. A double switch is done for congenitally corrected transposition of the great arteries. Anaesthetic management depends on repair and associated/residual lesions. Usually the repair is good, but late problems can arise from arrhythmias or stenosis/obstruction caused by sutures or conduits. Some operations include coronary re-implantation which might cause coronary perfusion problems. Maintain preload. Inotropes might be needed for ventricular function.

Keywords TGA, ccTGA, switch, atrial switch, arterial switch, ventricular switch, double switch · Jatene operation, Rastelli operation, Nikaidoh operation, REV, Mustard operation · Senning operation, Reparation a l'Etage Ventriculaire · Transposition of the great arteries, congenitally corrected transposition of the great arteries · Arrhythmia · Paediatric cardiac anaesthesia · Congenital cardiac lesion

Intra-atrial baffle (Senning or Mustard)

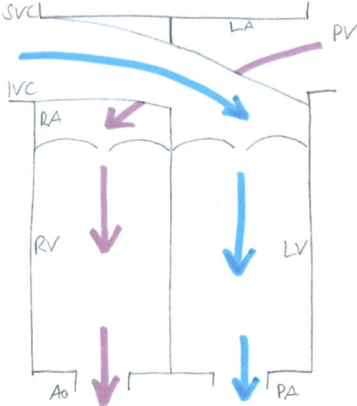

Fig. 43.1 Intra-atrial baffle to direct blood from the inferior and superior vena cava to the left atrium and ventricle, as well as directing blood from the pulmonary veins to the right atrium and ventricle

Ventricular switch (Rastelli)

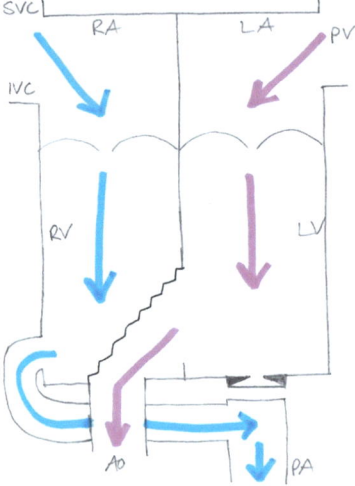

Fig. 43.2 Rastelli operation for transposition of the great arteries with pulmonary stenosis: an RV to PA conduit provides pulmonary blood flow

There are several options for repair of transposition of the great arteries, the decision is made by anatomy and associated lesions. In some cases the right ventricle is kept as the systemic ventricle, in some cases the operation makes the left ventricle the systemic one. Anaesthetic management depends on the details of the lesion and the repair.

43.1 Switch at Atrial Level: Mustard or Senning Procedure (Fig. 43.1)

- Creating an intra-atrial baffle to redirect systemic venous return to the left ventricle and pulmonary venous return to the right ventricle
- Senning: patient's own tissue used for baffle, Mustard: synthetic material
- Relatively high complication rate of arrhythmia, baffle obstruction and calcification, early right ventricular failure (systemic ventricle) → obsolete

Anaesthetic Management
- Right ventricle is systemic ventricle
- Echo:
    - ventricular function: might need inotropic support
    - baffle flow/obstruction/narrowing/kinking
    - additional lesions?
- Maintain preload
- Arrhythmias, both atrial and ventricular are common: have pacing equipment available
- Pacemaker?

43.2 Switch at Ventricular Level: Rastelli, REV, Nikaidoh (Fig. 43.2)

- Intraventricular manipulation of blood flow to respective arteries
- usually done in patients with transposition of the great arteries with ventricular septal defect and left ventricular (=pulmonary) outflow tract obstruction or double outlet right ventricle
    - Rastelli: Closure of ventricular septal defect to re-route blood from left ventricle to aorta, closure of pulmonary artery and connection of right ventricle to pulmonary artery via a conduit (unvalved or valved) or direct stitches with a patch
    - Reparation a l'Etage Ventriculaire = REV: excision of ventricular outlet septum, then closure of ventricular septal defect, reconstruction of right ventricular outflow tract without conduit
    - Nikaidoh: aortic root with coronaries is translocated posteriorly and left ventricular outflow tract is reconstructed with ventricular septal defect patch, right ventricular outflow tract is reconstructed with pericardial patch or homograft
- Damus-Kaye-Stansel operation (Chap. 37) can be done for ventricular septal defect and subaortic outflow tract obstruction

Anaesthetic Management
- Left ventricle is systemic ventricle
- Echo
  - Ventricular function/contractility
  - Obstruction/stenosis at anastomosis sites
  - Signs of coronary obstruction: angina, myocardial ischaemia/infarction with regional wall motion abnormalities, ventricular dysfunction?
  - Additional lesions?
- Arrhythmias due to suture lines in the ventricular septum→ have pacing available
- Rastelli: right ventricular outflow tract obstruction RVOTO can lead to high right ventricular pressures and subsequent right ventricular dysfunction and failure
- Might need inotropes, depending on biventricular function
- Maintain preload

43.3 Switch at Arterial Level: Jatene Operation (Fig. 43.3)

- Aorta and pulmonary artery are switched to their respective ventricles
- Coronaries get re-implanted into the neo-aorta

Anaesthetic Management
- Usually good repair with excellent long term outcomes
- Possible complications are stenosis/obstructions on anastomosis level and on the coronaries, resulting in myocardial ischaemia and ventricular dysfunction
- Echo for function, residual lesions, coronary blood flow, etc.

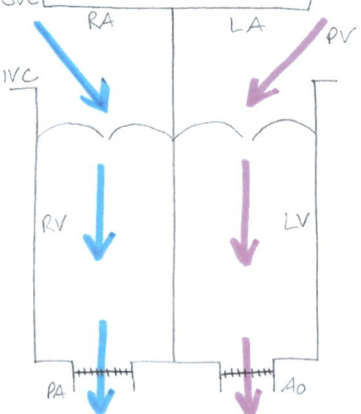

Fig. 43.3 Arterial switch (also called Jatene operation): attachment of pulmonary artery to right ventricle and aorta to left ventricle

43.4 Double Switch for Congenitally Corrected TGA

- Congenitally corrected transposition of the great arteries with additional problems like tricuspid regurgitation or right ventricular dysfunction need double switch repair, making the left ventricle the systemic ventricle
- consists of switch at atrial level combined with a switch at ventricular (Fig. 43.4) or arterial level (Fig. 43.5)
- presence of additional lesions like ventricular septal defects and/or pulmonary stenosis determine the type of switch (ventricular/arterial)

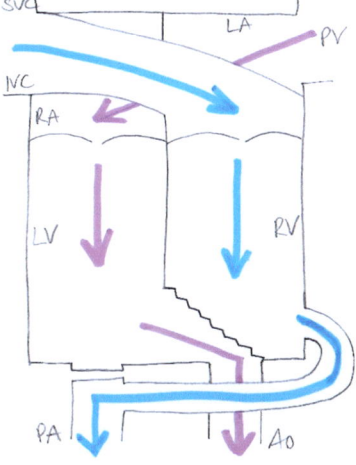

Fig. 43.4 Double switch with an intra-atrial baffle and an RV to PA conduit

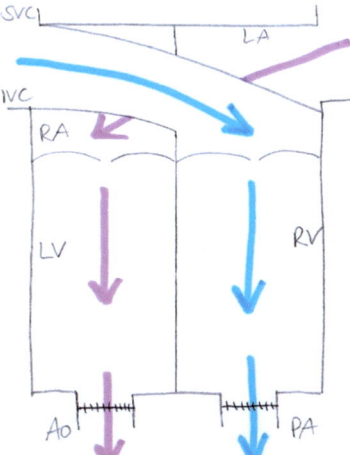

Fig. 43.5 Double switch operation: intra-atrial baffle and arterial switch

Anaesthetic Management
- multiple suture lines can create multiple obstructions → detailed Echo
- conduction problems: arrhythmias, both atrial and ventricular
- Pacemaker?
- possibility of deterioration of tricuspid regurgitation → promote forward flow by avoiding a decrease in heart rate, keep in sinus rhythm (pacing, if necessary)
- might need inotropes for myocardial impairment

Additional Reading

1. Christensen RE, Reynolds PI, Bukowski BK, Malviya S. Anaesthetic management and outcomes in patients with surgically corrected D-transposition of the great arteries undergoing non-cardiac surgery. Br J Anaesth. 2010;104(1):12–5.
2. Kiener A, Kelleman M, McCracken C, Kochilas L, Louis JD, Oster ME. Long-term survival after arterial versus atrial switch in d-transposition of the great arteries. Ann Thorac Surg. 2018;106(6):1827–33.
3. Deal BJ. Late arrhythmias after surgery for transposition of the great arteries. World J Pediatr Cong Heart Surg. 2011;2(1):32–6.
4. Görler H, Ono M, Thies A, Lunkewitz E, Westhoff-Bleck M, Haverich A, Breymann T, Boethig D. Long-term morbidity and quality of life after surgical repair of transposition of the great arteries: atrial versus arterial switch operation. Interact Cardiovasc Thorac Surg. 2011;12(4):569–74.

Chapter 44
Total Cavo-Pulmonary Connection (Fontan Operation)

Abstract A total cavo-pulmonary connection (TCPC) or Fontan circulation is the connection of the inferior vena cava to the pulmonary artery for a single ventricle lesion. This means pulmonary blood flow is dependent on pressure difference between central venous pressure and pulmonary vascular pressures. Any increase in intrapulmonary pressure or pulmonary vascular resistance decreases pulmonary blood flow and limits cardiac output.

Anaesthetic management aims to maintain cardiac output by maintaining central venous pressure with adequate preload and keeping pulmonary vascular resistance low with oxygen and mild hyperventilation.

Keywords Total cavo-pulmonary connection, TCPC · Fontan operation · Single ventricle, arrhythmia · Paediatric cardiac anaesthesia · Congenital cardiac lesion

A total cavo-pulmonary connection is the connection of inferior vena cava to pulmonary arteries with a conduit (Fig. 44.1), and together with the Glenn shunt (Chap. 38) means that both vena cavae, superior and inferior, are connected to the pulmonary circulation.

- Sometimes includes a fenestration into right atrium, which can shunt venous blood into the systemic circulation if pulmonary pressures are high
- Saturation in the low-mid 90's due to the fenestration (R → L shunt) and coronary venous return

Fig. 44.1 Total cavo-pulmonary connection: inferior vena cava is connected to the pulmonary artery

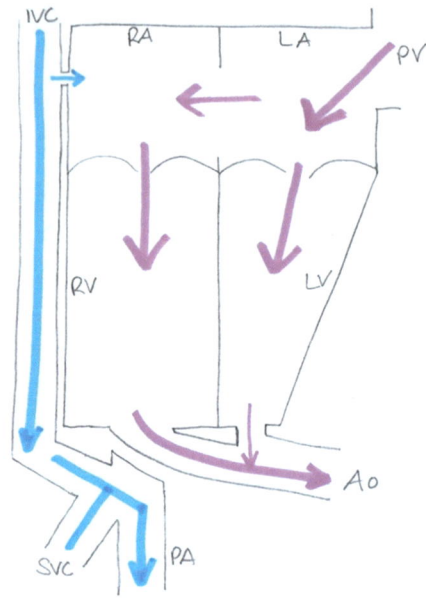

Fig. 44.2 Schematic of blood flow in a single ventricle circulation and its determinants

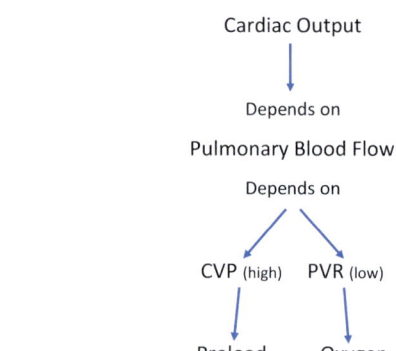

- Passive blood flow into the lungs (see Fig. 44.2):
  - Pulmonary blood flow dependent on pressure difference between central venous pressure and pulmonary vascular pressures
  - any increase in intrapulmonary pressure or pulmonary vascular resistance decreases pulmonary blood flow and limits cardiac output
  - hypovolaemia and low central venous pressure reduce pulmonary blood flow
- Sinus rhythm essential for good cardiac output

44 Total Cavo-Pulmonary Connection (Fontan Operation)

Complications in TCPC Patients: Immediate to Short Term
- Low cardiac output
  - Prone to thrombosis → usually on aspirin or warfarin
  - Limited ability to increase stroke volume
  - Limited ability to increase heart rate in response to exercise or sympathetic stimulation
- Arrhythmias, both atrial and ventricular

Long Term to Failing Circulation
- Venous congestion:
  - Increasing cyanosis due to R → L shunt via the fenestration
  - Lymphangiectasis, causing protein-losing enteropathy PLE
  - Persistent pleural effusions
  - Liver: dysfunction, hepatomegaly and cirrhosis
  - Reduced or dysfunctional clotting factors due to liver dysfunction and PLE
  - Ascites and peripheral oedema
- Increasing pulmonary vascular resistance over time due to:
  - Non-pulsatile blood flow results in pulmonary vascular disease
  - Development of aorto-pulmonary collaterals causing L → R shunt and increase in pulmonary vascular resistance
- Myocardial dysfunction: single ventricles are usually dilated, hypertrophic and hypo-contractile
- Arrhythmias, both atrial and ventricular

Anaesthetic Management
- Premedication
- Avoid dehydration, give IV fluids preoperatively, if necessary
- Might have difficult IV access (multiple operations, thrombosis)
- Positioning with elevated upper body, if possible, is advantageous for pulmonary blood flow
- Check clotting and have clotting products ready if needed
- Have pacemaker ready for external pacing and defibrillation if needed, arrhythmias are common, both atrial and ventricular
- Cardiac output is dependent on pulmonary blood flow, which is dependent on a relatively high central venous pressure and a low pulmonary vascular resistance
- If positive pressure ventilation:
  - $FiO_2 > 21\%$ to decrease pulmonary vascular resistance if needed
  - Avoid PEEP
  - Minimal peak inspiratory pressure
  - Short inspiration
- TCPCs might not respond well to inotropes, vasopressors and anti-arrhythmic medication

Additonal Reading

1. Veronese L, Swanevelder J, Brooks A. Anaesthesia for the child with a univentricular heart: a practical approach. S Afr J Anaesth Analg. 2021;27(3):144–53.
2. Christensen RE, Gholami AS, Reynolds PI, Malviya S. Anaesthetic management and outcomes after noncardiac surgery in patients with hypoplastic left heart syndrome: a retrospective review. Eur J Anaesthesiol. 2012 Sep;29(9):425–30.
3. Kerai S, Gaba P, Gupta L, Saxena K. Anaesthetic management of a child with unrepaired complete atrioventricular canal defect, double outlet ventricle and pulmonary stenosis for noncardiac surgery. Indian J Anaesth. 2022;66(18):342–S344.
4. Nicolson SC, Steven JM, Diaz LK, Andropoulos DB. Anesthesia for the patient with a single ventricle. In: Andropoulos DB, Stayer S, Mossad EB, Miller-Hance WC, editors. Anesthesia for congenital heart disease [Internet]. 1st ed. Wiley; 2015. p. 567–97. [cited 2024 Apr 27]. Available from: https://onlinelibrary.wiley.com/doi/10.1002/9781118768341.ch25.
5. Ribas Ball M, De Miguel Negro M, Galán Menéndez P, Dos Subirà L, Castro Alba MA, Martí Aguasca G. Anesthetic management of pulmonary artery banding in adult patient with single ventricle and uncorrected transposition of the great arteries. Revista Española de Anestesiología y Reanimación (English Edition). 2024 Feb;S2341192924000520.
6. Brown ML, DiNardo JA, Odegard KC. Patients with single ventricle physiology undergoing noncardiac surgery are at high risk for adverse events. Hammer G, editor. Pediatr Anesth. 2015;25(8):846–51.
7. Leyvi G, Wasnick JD. Single-ventricle patient: pathophysiology and anesthetic management. J Cardiothorac Vasc Anesth. 2010 Feb;24(1):121–30.
8. Greaney D, Honjo O, O'Leary JD. The single ventricle pathway in paediatrics for anaesthetists. BJA Educ. 2019;19(5):144–50.

Appendix

An Introduction to the Workbook

- These are 10 examples for practicing the anaesthetic planning for congenital cardiac lesions.
- Each example has an Echo first, then the explanation of the lesion and the anaesthetic planning.
- Read through the Echo.
- Draw the anatomy and the blood flow.
- List the pathophysiology of the lesion (Stenosis, Regurgitation, Shunt, Mixing).
- Consider the anaesthetic implications for each lesion:
  - Does the anaesthetic influence the lesion?
  - How does the lesion influence the anaesthetic?
- Compare your drawing/ considerations with the explanation provided.

cAVSD

Echo report:
Image quality: fair
SVC/ IVC to RA, PV to LA.
Large primum ASD and secundum defect with left to right flow.
Appearance of common AVV. Leaflets are mildly thickened. Moderate regurgitation. Difficult to visualise.
RV is dilated, good longitudinal function.
LV is small in size. Good systolic function. VSD very difficult to assess if unrestrictive. No obvious additional VSD's.

No RVOTO. The pulmonary valve opens well, trivial regurgitation. Confluent branch PA's. No obvious PDA flow seen.

No LVOTO. The aortic valve opens well, mildly increased flow (Vmax 2.0 m/s) however no significant OTO. Unable to assess aortic valve morphology.

Slender left sided aortic arch. No significant obstruction, Vmax 1.50 m/s. No diastolic decay. Pulsatile abdominal aorta.

Summary

- Balanced AVSD with a large atrial component and? small ventricular component.
- Appearance of common AVV with moderate regurgitation.

Anaesthetic Planning

Draw anatomy and blood flow (Fig. 1)

Pathophysiology

- L → R shunt at atrial and ventricular level
- One AV valve, regurgitant

Anaesthetic Considerations

- Limited anaesthetic influence on atrial shunt: maintain preload
- AV valve regurgitation: avoid bradycardia

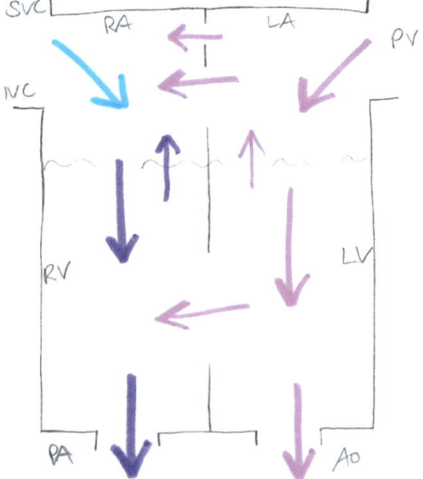

Fig. 1 complete AVSD with large ASD, small VSD and valve regurgitation

- Keep PVR high with FiO2 21% and mild hypoventilation
- Promote forward flow into Aorta with mild vasodilatation

DORV, Large VSD, Bilateral SVC

Echo Report
Image quality - good
Situs solitus, levocardia, AV-VA concordance.
The IVC/SVC drain into the RA, Left SVC demonstrated. Dilated Coronary sinus.
Fenestrated atrial septum with at least 2 atrial shunts demonstrated (left to right).
Normal sized RA.
Thin and mobile tricuspid valve, trivial regurgitation, weak signal for accurate RVSP estimation.
The right ventricle is normal in size, good parameters of longitudinal function. No evidence of RVH.
No RVOTO.
The pulmonary valve opens well, no stenosis, trivial regurgitation. Confluent branch PA's.
No obvious PDA flow seen.
Four pulmonary veins to the LA.
The LA appears mildly dilated visually.
Thin and mobile mitral valve, two papillary muscles demonstrated in usual position, trivial regurgitation
The left ventricle is normal in size. LV wall thickness appears within normal limits. Good radial and longitudinal systolic function. Unable to assess diastolic function accurately.
There is large outlet unrestrictive VSD with bidirectional flow and the aortic valve overrides the VSD >50%.
Tri-leaflet aortic valve opens well, no stenosis, no regurgitation. Normal coronary artery origins.
Left sided aortic arch with normal branching pattern.
Unobstructed aortic arch and proximal descending aorta. Pulsatile abdominal aorta.

Summary

DORV with large unrestrictive VSD.
Fenestrated atrial septum.
Preserved biventricular function.
Bilateral SVC noted.

Fig. 2 DORV, 2 ASDs, VSD with mixing

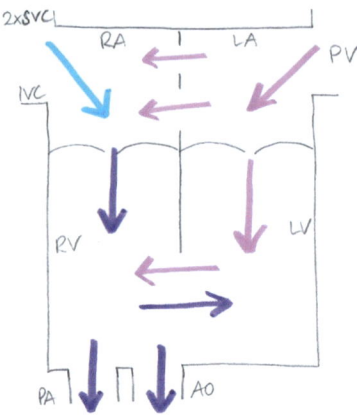

Anaesthetic Planning

Draw anatomy and blood flow (Fig. 2)

Pathophysiology

- L → R shunt at atrial level
- Bidirectional flow in VSD → mixing

Anaesthetic Considerations

- Chronic cyanosis causes polycythaemia
  - coagulation disorder, making them prone to both thrombosis and bleeding
  - risk of acute hyperviscosity syndrome when dehydrated
- → keep well hydrated, minimise fasting times, start IV fluids preoperatively if necessary
- Limited influence of anaesthesia on atrial shunt: maintain preload
- Maintain PVR and SVR at pre-op levels
  - Keep PVR high with FiO2 21% and mild hypoventilation
  - Avoid increase in SVR: promote forward flow into aorta with mild vasodilatation
- Change pressures as needed:
- → if saturation decreases: either increase SVR (fluids, vasopressors) or decrease PVR (oxygen, mild hyperventilation): this will redirect blood flow to the pulmonary circulation

Ebstein's Anomaly

Echo report:
Image quality: good
Situs solitus
Heart to the left with apex to the left
IVC and single right SVC to RA
4 pulmonary veins seen returning to the LA with normal doppler traces
Small Patent Foramen Ovale with left to right flow
Concordant AV-VA connections
Severe Ebstein's anomaly; small functional RV, the largest part is atrialised; the tricuspid valve is rotated anteriorly and to the left, severe TR
MV thin and mobile with unobstructed inflow and trivial regurgitation
Normal MV apparatus and two normal papillary muscles seen
The function of the LV appears impaired in the longitudinal plane and in particular the apex
Small apical muscular VSD with restricted left to right flow
AoV is tri-leaflet, thin and mobile with unobstructed outflow and no regurgitation
Two coronaries visualised arising from the usual anatomical position
PV thin and mobile with unobstructed outflow and trivial regurgitation
Confluent branch pulmonary arteries with laminar flow
Tiny PDA with left to right flow
Unobstructed aortic arch with laminar flow in the descending aorta and V max of 1 m/s
Pulsatile abdominal aorta
No pericardial effusion
No pleural effusion

Summary

1. Severe Ebstein's anomaly
2. Impaired LV function
3. Antegrade flow into the MPA
4. Tiny PDA with left to right flow
5. Small restricted apical muscular VSD
6. PFO with left to right flow

Anaesthetic Planning

Drawing of anatomy and blood flow (Fig. 3):
- small L → R shunts at atrial and ventricular level
- small L → R shunt at ductal level

Fig. 3 Ebstein's Anomaly with severely displaced tricuspid valve, small PFO, small VSD, tricuspid regurgitation and a patent ductus arteriosus

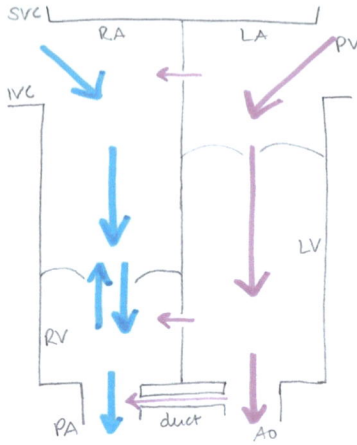

- tricuspid regurgitation
- impaired biventricular function

Anaesthetic Considerations

- Limited influence of anaesthesia on intracardiac L → R shunts
- Maintain preload. Careful with fluid boluses as the heart is already volume loaded.
- Ductal shunt not significant
- Problems: severe TR and impaired function:
  - Avoid bradycardia
  - Promote forward flow by reducing afterload:
  - FiO2 > 21% and mild hyperventilation for decreasing pulmonary blood pressure
  - Good anaesthesia and analgesia for systemic blood pressure
  - Might need inotropes for reduced function
  - No comment on RA size, but volume load from tricuspid regurgitation and anatomy can cause arrhythmia, both atrial and ventricular → have pacing equipment and medication available

Partial Anomalous Pulmonary Venous Drainage PAPVD

Echo report:
Image quality - good
Situs solitus, levocardia, AV-VA concordance.
SVC to RA. IVC to RA. Turbulent flow seen to enter IVC (Vmax 2.2 m/s, MPG 7 mmHg - Doppler consistent with pulmonary vein).

Superior secundum ASD with bidirectional flow.
Normal sized RA.
Thin and mobile tricuspid valve, mild to moderate regurgitation, estimated RVSP 85 mmHg + RAp
The right ventricle is severely dilated relative to LV. Preserved systolic function, RVS' 8 cm/s, TAPSE 13 mm.
The pulmonary valve opens well, no stenosis, trivial regurgitation. Confluent branch PA's. No obvious PDA flow seen.
2 left pulmonary veins seen to drain to LA. Unable to image right pulmonary venous drainage.
Normal sized LA.
Thin and mobile mitral valve, no regurgitation noted.
The left ventricle is normal in size. LV wall thickness is within normal limits. Septal flattening in systole consistent with pressure loaded RV. Good radial and longitudinal systolic function. Intact IVS with no obvious shunt.
No LVOTO.
Tri-leaflet aortic valve opens well, no stenosis, no regurgitation.
Normal coronary artery origins with prominent flow.
Unobstructed aortic arch and proximal descending aorta. Pulsatile abdominal aorta.

Summary

- Likely right pulmonary vein to IVC.
- Superior secundum ASD with bidirectional flow
- Severely dilated RV with preserved systolic function
- RVSP 86 mmHg + RAP, short PVAT and septal flattening in systole consistent with increased PAP.

Anaesthetic Planning

Drawing of anatomy and blood flow (Fig. 4):

Pathophysiology

- mild mixing on venous/ right side
- bidirectional shunt at atrial level
- tricuspid regurgitation

Anaesthetic Considerations

- No anaesthetic influence on the anatomy of pulmonary veins to right atrium

Fig. 4 partial anomalous pulmonary venous drainage with (likely right-sided) pulmonary vein draining into the inferior vena cava, a bidirectional shunt at atrial level, tricuspid regurgitation and a dilated right ventricle

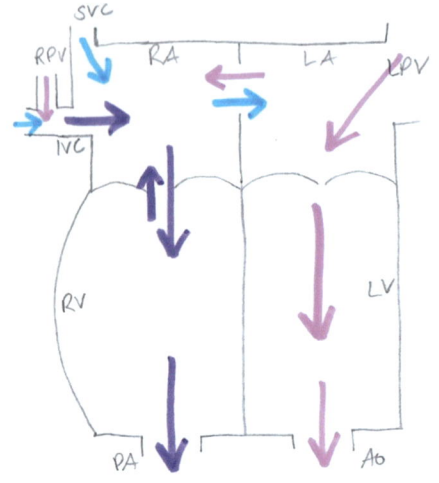

- Limited anaesthetic influence on atrial shunt: might have mild cyanosis
- Tricuspid regurgitation:
  - Avoid bradycardia
  - Promote forward flow: FiO2 > 21% and mild hyperventilation

- → severely dilated right ventricle: careful with fluid boluses
- → might need inotropes for function

Pulmonary Atresia, Intact Septum, Glenn Shunt

Echo report:
Good LV function
Unobstructed flow from SVC to branch PA's
IVC to atrial mass
Unobstructed pulmonary venous return to the LA
Unrestrictive interatrial communication
Hypoplastic RV and TV. Mild jet of TR with to and fro flow across the TV.
The MV has multiple chordae and prolapse of both leaflets resulting on multiple jets of several MR, along the zone of coaptation. Unobstructed inflow.
No LVOTO. Trileaflet AoV. Usual origin of the coronaries that appear of normal size.
Absent RV to PA connection (pulmonary atresia). Some flow noted in the MPA, filled retrogradely by the SVC.
No significant sinusoids identified
Imp: PA, IVS post Glenn

    Severe MR (as described above)
    Mild TR

Fig. 5 Pulmonary Atresia, unrestrictive ASD with R → L shunt, small right ventricle, tricuspid and mitral regurgitation. Glenn shunt: connection of superior vena cava to pulmonary artery

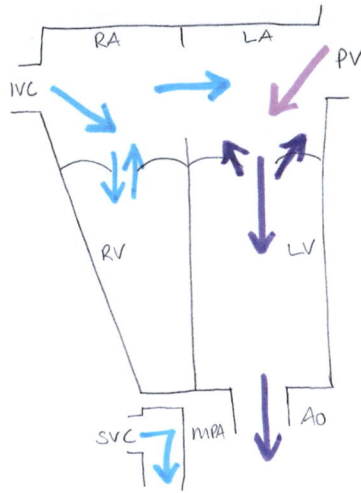

Anaesthetic Planning

Draw anatomy and blood flow (Fig. 5)

Pathophysiology

- R → L shunt at atrial level: needed for blood flow to left heart and systemic circulation
- Mixing
- Regurgitant valves

Anaesthetic Considerations

- Chronic cyanosis causes polycythaemia
    - coagulation disorder, making them prone to both thrombosis and bleeding
    - risk of acute hyperviscosity syndrome when dehydrated
- → keep well hydrated, minimise fasting times, start IV fluids preoperatively if necessary
- Regurgitation: avoid bradycardia
- Pulmonary blood flow dependent on SVC flow into lungs, dependent on high central venous pressure and low pulmonary vascular resistance
- FiO2 > 21% will not significantly improve saturations due to mixing
- mildly elevated upper body position might help with pulmonary blood flow
- Maintain preload, careful fluid management

Right Atrial Isomerism and Atrio-Ventricular Septal Defect

Echo:
Image quality: limited, patient upset.
Ao and IVC to the right of the midline. Right atrial isomerism.
Single IVC to RA
Single LSVC to left sided atrium
AVSD- large primum and significant secundum ASD
Pulmonary venous return join a confluence and drain to the left sided atrium, not normal but not obstructed at present
Normally related great arteries but side by side
Potential for subaortic obstruction, large infundibulum underneath the valve
Large VSD with slight aortic override
Common AV valve—more committed to the LV—mild AVVR
Tiny PDA
good sized branch PAs
Normal arch

Summary

Right atrial isomerism
Abnormal systemic venous return
AVSD with aortic override
Mild AVVR
Small PDA

Anaesthetic Planning

Drawing of anatomy and blood flow (Fig. 6):

- Mixing
- Several shunts, most likely bidirectional or L → R
- AV valve regurgitation

Anaesthetic Considerations

- Chronic cyanosis causes polycythaemia
  - coagulation disorder, making them prone to both thrombosis and bleeding
  - risk of acute hyperviscosity syndrome when dehydrated
- → keep well hydrated, minimise fasting times, start IV fluids preoperatively if necessary

Fig. 6 Right atrial isomerism with superior vena cava draining into the left sided atrium and inferior vena cava draining into the right sided atrium. Mixing on atrial and ventricular level, atrioventricular valve regurgitation. Patent ductus arteriosus

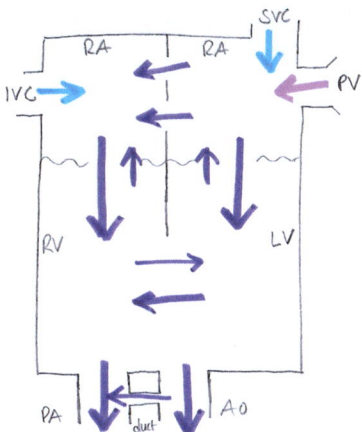

- Limited influence of anaesthesia on intracardiac shunts. Maintain preload
- Regurgitation: avoid bradycardia
- Maintain pulmonary and systemic vascular resistance at pre-op levels
  - PVR: FiO2 21%, mild hypoventilation
  - SVR: careful fluid management, haemodynamically stable anaesthesia
- In case of decrease in saturations: either decrease PVR with FiO2 > 21% and mild hyperventilation or increase SVR with fluids or a vasopressor
- → this will increase/ create a L → R shunt and more blood will be redirected to the pulmonary circulation.
- → if more blood is directed to pulmonary circulation, less blood will be going into systemic circulation, so blood pressure might drop

RAI, TAPVD, Bilateral SVCs, VSD, DORV, Malposed Great Arteries, PS

S/P TAPVD repair + BT shunt
S/P BT shunt take down and bilateral Glenn shunts
MPA still connected

Echo report:

Right and left pulmonary veins drain unobstructed to the left sided atrium
IVC drains to right sided atrium. SVC has been disconnected.
Common atrium.
The mitral valve is on the right and the tricuspid valve is on the left. Both AV valves are a similar size with normal inflows and very mild regurgitation.
There is a large VSD which is not committed to either outflow.

Both great arteries arise from the anterior, left sided, right ventricle. The aorta is anterior.

The aortic outflow is currently unobstructed. Although the VSD and the aortic valve can be seen in the same view (45), there is outlet septum which would need to be resected if the VSD were closed to the aorta (56).

The posterior pulmonary artery is much smaller and there is multilevel PS although forward flow across the pulmonary valve is still seen.

Conclusion

1. Dextrocardia, RAI, TAPVD, bilateral SVCs, VSD, DORV, malposed great arteries, PS.
2. Outlet septum would obstruct LV to Ao flow if the VSD were to be closed

Anaesthetic Planning

Draw anatomy and blood flow (Fig. 7).

Pathophysiology

- Mixing at atrial and ventricular level
- Mild regurgitation at AV valve level
- Severe pulmonary stenosis, blood flow into lungs via SVCs to pulmonary artery (Glenn shunt)

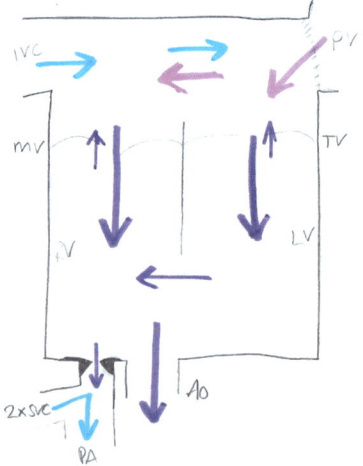

Fig. 7 Common atrium with mixing, bilateral superior vena cavae to pulmonary artery, inferior vena cava to common atrium. Pulmonary veins connected to the left sided atrium. Severe pulmonary stenosis

Anaesthetic Considerations

- Keep heart rate at pre-op level, avoid tachycardia, as the pulmonary stenosis is worse than the regurgitation at AV valve level
- Chronic cyanosis causes polycythaemia
  - coagulation disorder, making them prone to both thrombosis and bleeding
  - risk of acute hyperviscosity syndrome when dehydrated
- → keep well hydrated, minimise fasting times, start IV fluids preoperatively if necessary
- Dextrocardia impairs ST segment interpretation in standard ECG electrode positioning
- Pulmonary blood flow dependent on high central venous pressure and lower pulmonary vascular resistance
  - Maintain preload
  - Keep saturations at pre-op level
  - Elevating upper body might help with blood flow into lungs
  - FiO2 > 21% and mild hyperventilation will lower pulmonary vascular resistance

Tetralogy of Fallot's

Echo report:
Image quality—fair.
Situs solitus, levocardia, AV-VA concordance.
The IVC/SVC drain into the RA.
Small (central) atrial shunt, flow left-right across.
Normal RA size visually.
Thin and mobile tricuspid valve, trivial regurgitation, insufficient to estimate RVSP.
The right ventricle: good ventricular function, flow acceleration due to narrowing of the infundibulum, Vmax 3.6 m/s.
The pulmonary valve appears mildly dysplastic but the leaflets appear to have good acceleration, Vmax approx. 2.8 m/s, no significant pulmonary regurgitation.
Confluent branch PA's, of normal size.
Pulmonary veins to the LA.
Normal sized LA.
Thin and mobile mitral valve, no regurgitation noted.
The left ventricle appears non dilated, good LV systolic function.
Ventricular septal defect with aortic override—VSD appears unrestricted with bidirectional flow across. No addition VSDs demonstrated. Anterior deviation of conal septum demonstrated.
No LVOTO.

Tri-leaflet aortic valve opens well, no stenosis, no regurgitation. RCA origin is slightly clockwise rotated, LCA appears to arise in usual position.
Unobstructed aortic arch and proximal descending aorta. Pulsatile abdominal aorta.

Summary

- Large VSD (Fallot type)
- Infundibular pulmonary stenosis, Vmax 3.6 m/s
- Well developed confluent branch PA's
- Atrial shunt, flow L-R

Anaesthetic Planning

Drawing of anatomy and blood flow (Fig. 8)

- L → R shunt at atrial level
- bidirectional shunt/ mixing at ventricular level
- moderate right ventricular outflow tract obstruction RVOTO

Anaesthetic Considerations

- no anaesthetic influence on atrial shunt
- Chronic cyanosis causes polycythaemia
    - coagulation disorder, making them prone to both thrombosis and bleeding
    - risk of acute hyperviscosity syndrome when dehydrated

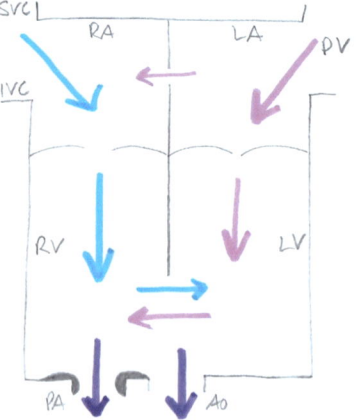

Fig. 8 Small atrial L → R shunt, bidirectional ventricular shunt with mixing, pulmonary stenosis

- → keep well hydrated, minimise fasting times, start IV fluids preoperatively if necessary
- minimal anaesthetic influence on ventricular shunt/ mixing:
  - promote forward flow through RVOT with FiO2 > 21% and mild hyperventilation
  - Careful fluid management: redirecting blood flow to the lungs will create relative hypovolaemia in the systemic circulation with lower blood pressure.
  - pulmonary stenosis: avoid tachycardia

Repaired Truncus Arteriosus with Subaortic Stenosis and Regurgitation

Echo Report:
Image quality is sub-optimal due to body habitus
The left ventricle is non dilated by absolute measurements, but looks globular from apical views. The LV systolic function is visually normal, but in the presence of significant truncal valve regurgitation there might be a degree of impairment.
The IVS looks thickened, although difficult to measure it accurately.
Normal size LA. No residual flow across the ventricular septum.
The mitral valve posterior leaflet appears to be short, but both leaflets open well. There is mild posteriorly directed mitral regurgitation.
The truncal valve leaflets not seen clearly, but from views obtained appear to open well with an eccentric jet of severe aortic regurgitation, holodiastolic flow reversal seen in the aortic arch and abdominal aorta. There is an impression of subvalvar membrane (best seen from apical views), causing flow acceleration starting at that region with LVOT/truncal valve Vmax = 2.5 m/s.
Unobstructed aortic arch and proximal descending aorta, no diastolic decay. Pulsatile abdominal aorta.
Holodiastolic reverse flow seen both in proximal descending and abdominal part of aorta.
The right ventricle is visually normal in size with impression of preserved radial function, but mildly reduced longitudinal markers, TAPSE 16 mm.
Normal size RA.
The tricuspid valve is thin and mobile, trivial TR detected. Estimated RVSP =40 mmHg + RAP
Pulmonary conduit in situ, peak velocity 3.2 m/s, with trivial PR. Branch PAs appear slender, stents seen.

Conclusion

- Severe AR of the truncal valve, holodiastolic reverse flow seen in aortic arch and abdominal aorta.
- Sub-valvar membrane with LVOT/truncal valve Vmax =2.5 m/s.
- Pulmonary conduit in situ, peak velocity 3.2 m/s, trivial PR. Stents in branch PAs.
- Trivial TR, estimated RVSP =40 mmHg + RAP
- Normal size RV with mildly reduced longitudinal function
- Visually globular LV with possibly a degree of impairment due to significant truncal valve regurgitation. Visually LVH.

Anaesthetic Planning

Draw anatomy and blood flow (Fig. 9):

Pathophysiology

- Combined aortic/truncal stenosis (subvalvar membrane) and regurgitation
- Impairment of LV function, mild impairment of RV function

Anaesthetic Considerations

- **High risk anaesthesia**
- Keep heart rate at pre-op level
- Improve forward flow by mildly reducing systemic vascular resistance: good anaesthesia and analgesia

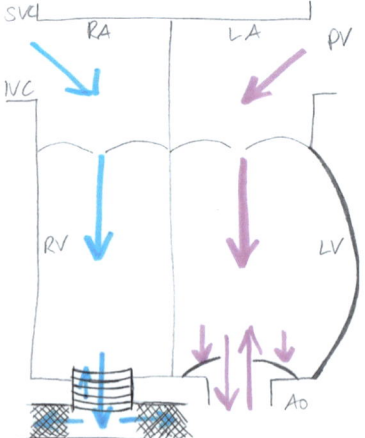

Fig. 9 Pulmonary valve homograft with mild regurgitation and stents in left and right pulmonary arteries. Suboartic membrane with obstruction and severe regurgitation. Left ventricular hypertrophy and dilatation

- Maintain perfusion pressure in coronaries
- Careful fluid management
- might benefit from inotropes (Milrinone: inotropic, can be given peripherally, doesn't increase O2 demand of the heart, vasodilatation)

Ventricular Septal Defect

Echo report:
Situs solitus, levocardia, (AVVA concordance)
IVC/SVC to the RA. Pulmonary veins to the LA.
Small atrial shunt in the secundum region. No evidence of a primum defect.
Normal sized RA, mildly dilated LA.
There are two separate junctions and two separate AV valves with tricuspid and mitral valve morphology.
Mild central MR.
Mild to moderate TR, PPD 84 mmHg, suggestive of elevated RV pressure/PVR resistance still raised.
The right ventricle is normal in size (RVD1 21 mm), normal parameters of longitudinal function, TAPSE 10 mm. Mild RVH noted visually.
The left ventricle is normal in size (LVIDd 22 mm). LV wall thickness is within normal limits visually. Good/normal radial and longitudinal systolic function.
Large outlet VSD with bidirectional unrestrictive flow. The VSD extends towards the inlet region. No additional VSDs seen.
No RVOTO.
Normal appearance of the LVOT with no obstruction (Vmax 0.7 m/s)
The pulmonary valve is normal (thin and mobile), trivial regurgitation. Confluent branch PAs. LPA 5 mm, RPA 5 mm.
Small PDA flow seen, low velocity ~ 1.5–2.0 m/s, predominately left to right.
Tri-leaflet aortic valve opens well, no stenosis, trace regurgitation.
Coronary arteries were difficult to image, the left has normal origin, the right was difficult but appears to arise in the usual position.
Unobstructed aortic arch and proximal descending aorta. Arch sidedness not assessed. Previously reported as left sided.
Pulsatile abdominal aorta.

Summary

Large unrestrictive outlet VSD.
Small atrial shunt.
PDA.
Preserved biventricular function

Fig. 10 Small L → R shunt at atrial and ductal level. Large ventricular bidirectional shunt with mixing. Tricuspid regurgitation

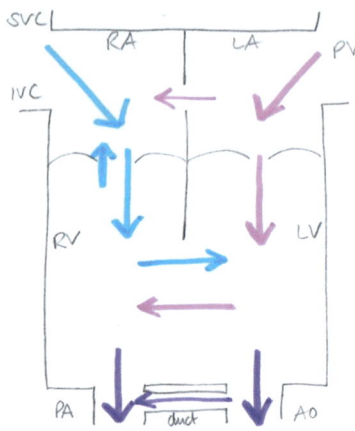

Anaesthetic Planning

Drawing of anatomy and blood flow (Fig. 10).

Pathophysiology

- small atrial L → R shunt
- tricuspid regurgitation
- large ventricular shunt/ mixing
- small ductal L → R shunt

Anaesthetic Considerations

- Limited anaesthetic influence on atrial shunt: maintain pre-load
- Tricuspid regurgitation: Avoid bradycardia
- Ventricular shunt/ mixing:
  - Oxygen and mild hyperventilation will promote forward flow into pulmonary circulation and might influence the bidirectional shunt to become L → R
  - Avoid increase in SVR, but ensure adequate systemic blood pressure for perfusion of coronaries.
- Ductal L → R shunt: avoid decrease in PVR, avoid increase in SVR

Index

A

ALCAPA, *see* Anomalous left coronary artery from pulmonary artery
Anesthetic planning, 2, 3
Anomalous left coronary artery from pulmonary artery, 25–27
Anomalous left pulmonary artery, 30
Antenatal circulation, 12
Aortic arch abnormalities, 9, 29–32
Aortic arch, interrupted, 33–36
Aortic valve regurgitation, 39
Aortic valve stenosis, 37–39
Aorto-pulmonary septal defect, *see* Aorto-pulmonary window
Aorto-pulmonary window, 41, 42
A-P window, *see* Aorto-pulmonary window
Arch hypoplasia, *see* Aortic arch abnormalities
Arterio-venous fistulae, *see* Arterio-venous malformation
Arterio-venous malformation, 18, 48–51, 161
ASD, *see* Atrial septal defect
Asplenia syndrome, *see* Right atrial isomerism
Atrial septal defect, 3, 4, 17, 43–45, 53, 57, 67, 68, 80–82, 86, 90, 103, 104, 107, 125, 126, 135, 136, 139, 172, 173
Atrial septostomy, 163
Atrio-ventricular septal defects, 19
  complete, 53–56
  partial, 57, 58
AVM, *see* Arterio-venous malformation
AVSD, *see* Atrio-ventricular septal defect

B

Balloon atrial septostomy, 94, 104, 135, 172–173
BAV, *see* Bicuspid aortic valve
Bicuspid aortic valve, 9, 33, 36, 37, 63, 64
Blalock-Taussig shunt, 74, 80, 84, 88, 92, 95, 105, 127, 154, 157, 158, 167, 168
BT shunt. See Blalock-Taussig Shunt
Bulbo-ventricular foramen, 71–74

C

ccTGA, *see* Congenitally Corrected transposition of the great arteries
Coarctation, 16, 29, 33, 63–66, 71, 93, 125
Congenitally corrected transposition of the great arteries, 8, 59–62, 71, 183
Cooley shunt, 154
Cord clamping, 12
Coronary sinusoids, 103, 105
Cor triatriatum, 67–69

D

Damus-Kaye-Stansel, 74, 127, 141, 157, 158, 181
DiGeorge, 33, 36, 143
DILV, *see* Double inlet left ventricle
DORV, *see* Double outlet right ventricle
Double aortic arch, 29, 30
Double inlet left ventricle, 8, 19, 71–74, 160

Double outlet right ventricle (DORV), 19, 54, 77–80, 86, 88, 118, 160, 172, 181
Doubly committed VSD, 77, 80
Duct dependent, 4, 82, 121, 122, 140
Ductal stent, 74, 105, 163–165, 173–174
Duct-dependent, 2, 9, 17, 35, 38, 83, 87, 91, 93–95, 103, 104, 121, 125, 126, 173
Ductus arteriosus, 11–13, 63, 111, 173

E
Ebstein's anomaly, 81–84, 131, 160
Eisenmenger syndrome, 18, 44
Ejection fraction, 9
Epsilon aminocaproic acid, 21

F
Foramen ovale, 11, 12, 16, 82, 113, 129, 130
Fractional area change, 9
Fractional shortening, 9

G
Glenn shunt, 56, 62, 74, 80, 84, 88, 92, 95, 105, 127, 137, 159–161, 167, 185

H
Hemifontan, 127, 159, 160
Hemitruncus, 143, 145
Hereditary haemorrhagic telangiectasia, 49
Heterotaxy syndrome, 8, 85–92, 135, 137
HLHS. *See* Hypoplastic left heart syndrome, 93
Hybrid procedure, 95, 163–165
Hyperoxia test, 2
Hypertrophic cardiomyopathy, 98, 99
Hyperviscosity syndrome, 19–22, 55, 61, 72, 79, 84, 87, 91, 94, 104, 119, 122, 127, 130, 131, 136, 140, 146, 158, 161, 164, 169, 172
Hypoplastic aortic arch, 33, 63–66, 93, 167
Hypoplastic left heart syndrome (HLHS), 65, 93–95, 160, 163–165
Hypoplastic right heart, 125–127

I
Interrupted aortic arch, *see* Aortic arch, interrupted
Intra-atrial baffle, 160, 180, 181, 183

J
Jatene operation, 182–183

K
Konno operation, 38, 176, 177

L
Laks shunt, 154, 155
LCA, *see* Left coronary artery
LCOS, *see* Low cardiac output syndrome
Left atrial isomerism, 85–88
Left coronary artery, 25–27
Left ventricular backwards failure, 16
Low cardiac output failure, 99
Low cardiac output syndrome (LCOS), 16–18, 38, 39, 67, 94, 99, 104, 136

M
Major aorto-pulmonary collateral arteries, 121–123
MAPCA, *see* Major aorto-pulmonary collateral arteries, 122
Melbourne shunt, 154
Mitral valve prolapse, 100–101
Mitral valve regurgitation, 99–100
Mitral valve stenosis, 97–98
Mixing lesions, 8, 19, 20, 36, 53, 55, 72, 74, 94, 95, 157, 161, 164, 169
Mustard procedure, 180, 181
Myocardial performance index, 9

N
Nikaidoh, 141, 181–182
Norwood operation, 95, 163, 167–169

O
Obstructive lesions, 16, 98, 140

P
PAPVD, *see* Partial anomalous pulmonary venous drainage
Partial anomalous pulmonary venous drainage, 18, 86, 107–109
Patent ductus arteriosus, 111, 112, 135
Patent foramen ovale, 81, 103, 109, 114, 130, 131, 135

Index

PDA, *see* Patent ductus arteriosus
Persistent ductus arteriosus, 18
Pink Fallot, *see* Tetralogy of Fallot
Polycythaemia, 19, 49
Polysplenia syndrome, *see* Left atrial isomerism
Potts shunt, 154, 156
Prostaglandin, 2, 17, 35, 38, 80, 83, 87, 91, 95, 104, 114, 140, 173
Prostin, 35, 38, 83, 91, 114
Pseudocoarctation, 64
Pseudotruncus, 145
Pulmonary artery band, 56, 61, 74, 80, 88, 127, 163–165, 171–172
Pulmonary atresia, 87, 90, 91, 103–105, 118, 121–123, 145, 173, 174
   with VSD (*see* Tetralogy of Fallot's with Pulmonary Atresia)
Pulmonary hypertension, 2, 15, 16, 18, 39–42, 44, 50, 53, 57, 64, 65, 67, 69, 74, 79, 98, 99, 109, 111, 126, 131, 143, 145, 147–149, 154, 163, 171
Pulmonary hypertensive crisis, 2
pulmonary regurgitation, *see* Pulmonary valve regurgitation
Pulmonary valve regurgitation, 115, 119
Pulmonary valve stenosis, 113, 114

R

Rashkind procedure, *see* Balloon atrial septostomy, 94
Rastelli, 80, 141, 180–182
RCA, *see* Right coronary artery
Regurgitant lesions, 15, 99, 116
Reparation a l'Etage Ventriculaire, 141, 181–182
REV, *see* Reparation a l'Etage Ventriculaire
Reverse Potts shunt, *see* Potts shunt
Rheumatic fever, 39, 97, 129, 131
Right aortic arch, 29, 31, 118, 143
Right atrial isomerism, 89–92, 135
Right coronary artery, 25, 50
Right ventricular fractional area change, 9
Right ventricular outflow tract obstruction, 117, 118, 182
Right ventricular outflow tract (RVOT) stent, 120
Ross operation, 38, 175, 176
RV fractional area change, 9
RVOTO, *see* Right ventricular outflow tract obstruction

S

SAM, *see* Systolic anterior motion of mitral valve
Sano shunt, 157, 158, 167, 168
SCD, *see* Sudden cardiac death
Scimitar syndrome, 108
Senning procedure, 181
Sequential segmental analysis, 7, 8
Shone's complex, 37
Shunt lesions, 17, 18, 153
Single ventricle pathway, 56, 62, 80, 84, 88, 105, 127, 137
Sinus venosus, *see* Atrial septal defect
Sub-aortic VSD, 77–80
Subclavian flap, 65
Subclavian steal syndrome, 65
Sub-pulmonary VSD, 77–79
Sudden cardiac death, 26, 38, 100
Supraventricular tachycardia (SVT), 60, 81–83, 100
Switch operations, 180–184
Systolic anterior motion of mitral valve, 98, 99

T

TAPSE, *see* Tricuspid annular plane systolic excursion
TAPVD, *see* Total anomalous pulmonary venous drainage
Taussig-Bing anomaly, 63, 77
Tei index, *see* Myocardial performance index
Tetralogy of Fallot, 54–56, 72, 79, 115, 117–123, 145
Total anomalous pulmonary venous drainage, 18, 90, 91, 134–137, 172
Total cavo-pulmonary connection, 56, 62, 74, 80, 84, 88, 92, 95, 105, 127, 137, 185, 187
Tracheomalacia, 31, 32, 122, 123
Tranexamic acid, 21
Transposition of the great arteries, 8, 54, 71, 73, 86, 90, 91, 125, 139–141, 158, 172, 180, 181
Tricuspid annular plane systolic excursion, 9
Tricuspid atresia, 160
Tricuspid regurgitation, 183
Tricuspid stenosis, 83
Tricuspid valve atresia, 3, 4, 19, 125–127, 172, 173
Tricuspid valve regurgitation, 2, 15, 42, 44, 45, 60–62, 79, 81–83, 98, 109, 114, 115, 131, 148, 149

Tricuspid valve stenosis, 83, 129–130
Trisomy, 53
Truncus arteriosus, 19, 36, 41, 143–146

U
Unifocalisation, 123
Unroofed coronary sinus, *see* Atrial septal defect

V
Vascular rings, *see* Aortic arch abnormalities
Vascular slings, 31

Vein of Galen, 49
Ventricular septal defect, 1, 3, 4, 15, 17, 33, 36, 53, 57, 60–62, 72, 77, 79, 80, 86–88, 94, 117–119, 121–123, 125–127, 135, 139, 141, 143, 145–149, 181, 183
VSD, *see* Ventricular septal defect

W
Waterston shunt, 154, 156
Williams syndrome, 37
Wolff-Parkinson-White syndrome, 82

MIX
Papier aus verantwortungsvollen Quellen
Paper from responsible sources
FSC® C105338

If you have any concerns about our products,
you can contact us on
ProductSafety@springernature.com

In case Publisher is established outside the EU,
the EU authorized representative is:
**Springer Nature Customer Service Center GmbH
Europaplatz 3, 69115 Heidelberg, Germany**

Printed by Libri Plureos GmbH
in Hamburg, Germany